Hydraulic and Hydrologic Engineering

T0305698

Hydraulic and Hydrologic Engineering
Fundamentals and Applications

Zohrab A. Samani

CRC Press
Taylor & Francis Group
Boca Raton London New York

CRC Press is an imprint of the
Taylor & Francis Group, an **informa** business

First edition published 2022
by CRC Press
6000 Broken Sound Parkway NW, Suite 300, Boca Raton, FL 33487-2742

and by CRC Press
2 Park Square, Milton Park, Abingdon, Oxon, OX14 4RN

CRC Press is an imprint of Taylor & Francis Group, LLC

© 2022 Zohrab A. Samani

Reasonable efforts have been made to publish reliable data and information, but the author and publisher cannot assume responsibility for the validity of all materials or the consequences of their use. The authors and publishers have attempted to trace the copyright holders of all material reproduced in this publication and apologize to copyright holders if permission to publish in this form has not been obtained. If any copyright material has not been acknowledged please write and let us know so we may rectify in any future reprint.

Except as permitted under U.S. Copyright Law, no part of this book may be reprinted, reproduced, transmitted, or utilized in any form by any electronic, mechanical, or other means, now known or hereafter invented, including photocopying, microfilming, and recording, or in any information storage or retrieval system, without written permission from the publishers.

For permission to photocopy or use material electronically from this work, access www.copyright.com or contact the Copyright Clearance Center, Inc. (CCC), 222 Rosewood Drive, Danvers, MA 01923, 978-750-8400. For works that are not available on CCC please contact mpkbookspermissions@tandf.co.uk

Trademark notice: Product or corporate names may be trademarks or registered trademarks and are used only for identification and explanation without intent to infringe.

ISBN: 978-1-032-26278-9 (hbk)
ISBN: 978-1-032-26284-0 (pbk)
ISBN: 978-1-003-28753-7 (ebk)

DOI: 10.1201/9781003287537

Typeset in Times
by SPi Technologies India Pvt Ltd (Straive)

Contents

List of Figures

List of Tables

Author's Biography

Prof. Zohrab Samani is Endowed Professor of Water Resources in the Civil Engineering Department at New Mexico State University, Las Cruces, NM, USA. Prof. Samani has more than 40 years of experience in research, teaching, consulting and international technology transfer. His areas of expertise include remote sensing technology, water resources development, groundwater hydrology and modeling, climate change impact on water resources, open-channel hydraulics, pipe network design, pumping systems design and management, well design and development, and renewable energy and sustainable resource development. He has received numerous awards for teaching and research, including Distinguished Educator Award (New Mexico PNM Foundation), ASCE Best Paper Awards, Outstanding Researcher Award (New Mexico State University), Globalization Award (Utah State University and New Mexico State University), and the Outstanding Faculty Award (New Mexico State University), Foreman Professorship and Bromilow outstanding researcher award.

1 Pipeline Design

1.1 DIMENSIONS ASSOCIATED WITH COMMON PHYSICAL QUANTITIES

TABLE 1.1
Standard Dimensions

	FLT System	MLT System
Acceleration	LT^{-2}	LT^{-2}
Angle	$F°L°T°$	$M°L°T°$
Angular acceleration	T^{-2}	T^{-2}
Angular velocity	T^{-1}	T^{-1}
Area	L^2	L^2
Density	$FL^{-4}T^2$	ML^{-3}
Energy	FL	ML^2T^{-2}
Force	F	MLT^{-2}
Frequency	T^{-1}	T^{-1}
Length	L	L
Mass	$FL^{-1}T^2$	M
Modulus of elasticity	FL^{-2}	$ML^{-1}T^{-2}$
Moment of a force	FL	ML^2T^{-2}
Moment of inertia (area)	L^4	L^4
Moment of inertia (mass)	FLT^2	ML^2
Momentum	FT	MLT^{-1}
Power	FLT^{-1}	ML^2T^{-3}
Pressure	FL^{-2}	$ML^{-1}T^{-2}$
Specific heat	$L^2T^{-2}\Theta^{-1}$	$L^2T^{-2}\Theta^{-1}$
Specific weight	FL^{-3}	$ML^{-2}T^{-2}$
Strain	$F°L°T°$	$M°L°T°$
Stress	FL^{-2}	$ML^{-1}T^{-2}$
Surface tension	FL^{-1}	MT^{-2}
Temperature	Θ	Θ
Time	T	T
Torque	FL	ML^2T^{-2}
Velocity	LT^{-1}	LT^{-1}
Dynamic viscosity	$FL^{-2}T$	$ML^{-1}T^{-1}$
Kinematic viscosity	L^2T^{-1}	L^2T^{-1}
Volume	L^3	L^3
Work	FL	ML^2T^{-2}
PSI	FL^{-2}	$ML^{-1}T^{-2}$

DOI: 10.1201/9781003287537-1

1.2   SUMMARY OF EQUATIONS

This chapter describes the basic principles of fluid mechanics/hydraulics, of which there are three:

Conservation of energy
Conservation of mass
Conservation of forces/momentum principles

Conservation of energy, as applied to hydraulic systems, can be summarized as:

$$E_1 = E_2 + h_L - h_p + h_t$$

where E_1 is upstream energy, E_2 is downstream energy, h_L is the head loss, h_p is the head of any pump in the system and h_t is the head associated with any turbine or other energy-consuming device in the system. If we do not have a pump or turbine, then the conservation of energy equation will simplify to:

$$\frac{V_1^2}{2g} + \frac{P_1}{\gamma} + Z_1 = \frac{V_2^2}{2g} + \frac{P_2}{\gamma} + Z_2 + h_L$$

where $h_L = h_f$ (friction loss) + h_m (minor loss)
    Conservation of mass is described as:

$$\rho_1 A_1 V_1 = \rho_2 A_2 V_2$$

where ρ_1 and ρ_2 are densities, A_1 and A_2 are areas and V_1 and V_2 are flow velocities.
    If the density of the fluid is constant, then the conservation of mass simplifies to:

$$A_1 V_1 = A_2 V_2 = Q$$

where Q is the volumetric flow rate.
    Conservation of forces is based on Newton's second law of motion and can be expressed as:

$$\Sigma \vec{F} = \rho Q \left(\vec{V_2} - \vec{V_1} \right)$$

where F, V_1 and V_2 are vector force and velocities respectively, ρ is the density and Q is the volumetric flow rate.

1.2.1   EMPIRICAL EQUATIONS

The Hazen–William (H–W) equation is an empirical closed-form solution of Equation (1.2) below. It is limited to $V < 3$ m/s and diameter > 2 inches

$$V = 1.318 C_{\text{HW}} R_{\text{h}}^{0.63} S^{0.54} \text{ English system}$$

$$V = 0.85 C_{\text{HW}} R_{\text{h}}^{0.63} S^{0.54} \text{ Metric} (\text{SI}) \text{system}$$

Some conventional forms of the H–W equation are shown below:

$$h_{\text{f}} = 10.5 (L) \times \left(\frac{Q}{C} \right)^{1.852} (D)^{-4.87}$$

h_{f} = friction loss, ft
L = length, ft
Q = flow rate, gpm
D = diameter, inches
C = Hazen–William coefficient

and

$$Q = 0.28 C \times (D^{2.63})(S^{0.54})$$

Q = flow rate, gpm
D = diameter, inches
C = Hazen–William coefficient
S = slope of energy grade line, dimensionless

1.3 PRINCIPLES OF HYDRAULICS

1.3.1 CONSERVATION OF ENERGY

Conservation of energy constrains the energy components in a fluid and the interchange of those components as the fluid moves. In a fluid, energy takes three forms: gravitational energy, kinetic energy (in the form of velocity) and pressure. The conservation of energy equation is shown below:

$$E_1 = E_2 + h_{\text{L}} - h_{\text{p}} + h_{\text{t}}, \tag{1.1}$$

where E_1 is upstream energy, E_2 is downstream energy, h_{L} is the head loss, h_{p} is the head (energy) produced by any pump in the system and h_{t} is head (energy) taken out by any turbine in the system. If we do not have a pump or turbine, then the conservation of energy equation will simplify to:

$$E_1 = E_2 + h_{\text{L}}$$

The components of energy are:

$$E = \frac{v^2}{2g} + \frac{p}{\gamma} + z$$

where E is energy and v, p, and z are the velocity, pressure and gravitational energy respectively. The above equation is also called the Bernoulli equation.

Z, the gravitational energy, is the distance measured vertically from a horizontal reference plane. Substituting the energy components into Equation (1.1), the conservation of energy equation will change to:

$$\frac{V_1^2}{2g} + \frac{P_1}{\gamma} + Z_1 = \frac{V_2^2}{2g} + \frac{P_2}{\gamma} + Z_2 + h_L \tag{1.2}$$

where $h_L = h_f$ (friction loss) $+ h_m$ (minor losses)

The term pipe flow usually refers to a full pipe.

The pressure in a pipe generally varies from one point to another, but a mean value is normally used over a particular section of interest. In other words, the regional pressure variation in a given cross section is commonly neglected unless otherwise specified.

In most engineering computations the section mean velocity V, defined as the discharge, Q, divided by the cross-sectional area, A, is used as shown below:

$$V = \frac{Q}{A}, \tag{1.3}$$

1.3.2 CONSERVATION OF MASS (CONTINUITY)

Conservation of mass is described as:

$$Q_1 = Q_2$$

or

$$\rho_1 A_1 V_1 = \rho_2 A_2 V_2$$

where ρ is the density of fluid, A is the cross-sectional area of flow and V is the velocity. If the density does not change, as in the case of water which is relatively incompressible, then the equation of continuity will simplify to:

$$A_1.V_1 = A_2.V_2 \tag{1.4}$$

Example 1.1

Water is flowing from a pipe with diameter of 8 inches and velocity of 3 ft/s into a pipe with diameter of 4 inches. What would be the velocity of water in the second pipe?

Solution:

$$A_1.V_1 = A_2.V_2$$

$$\frac{\pi(8)^2}{4}(3) = \frac{\pi(4)^2}{4}V_2$$

and V_2 is calculated as 12 ft/s.

1.3.3 Conservation of Forces

Conservation of forces in hydraulics is defined as:

$$\sum \vec{F} = \rho Q \left(\vec{V_2} - \vec{V_1} \right) \tag{1.5}$$

in which F is vector force, ρ is density and V is vector velocity. This equation is based on Newton's second law of motion which is more commonly applied to objects in motion. Examples of this principle are the phenomena of water hammer and hydraulic jump.

1.4 REYNOLDS NUMBER

In the nineteenth century the Irish engineer Osborne Reynolds performed a series of experiments in pipe flow. Reynolds developed a dimensionless parameter which is used in fluid mechanics to determine the level of turbulence in pipe flow. Reynolds divided the flow in pipes into two categories: (1) laminar flow and (2) turbulent flow.

In laminar flow, the particles of the fluid move in smooth paths, while in turbulent flow the particles move in a random fashion. Reynolds defined the dimensionless Reynolds number as:

$$\text{Re} = \frac{\rho DV}{\mu} \qquad (1.6)$$

in which ρ is density of the fluid, D is the diameter of the pipe, V is the velocity and μ is the dynamic viscosity of the fluid.

Kinematic viscosity in fluids is defined as the dynamic viscosity divided by the density of the fluid:

$$\upsilon = \frac{\mu}{\rho} \qquad (1.7)$$

Substituting kinematic viscosity Equation (1.7) into Equation (1.6) results in:

$$\text{Re} = \frac{DV}{\upsilon} \qquad (1.8)$$

Reynolds numbers less than or equal to 2000 represent laminar flow, while Reynolds numbers more than 4000 represent turbulent flow. The Reynolds value between 2000 and 4000 is called the critical zone (Figure 1.1). For the critical zone, the friction factor can be calculated using laminar flow line.

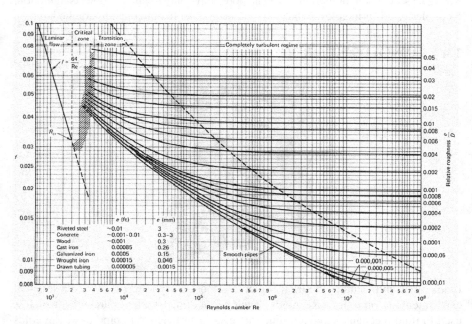

FIGURE 1.1   Moody diagrams. Friction loss is calculated as function of e/D and Re.

TABLE 1.2
Viscosity of Water at Various Temperatures

Temperature	Dynamic Viscosity		Kinematic Viscosity	
°C	N-s/m² × 10⁻³	lb-s/ft² × 10⁻⁵	m²/s × 10⁻⁶	ft²/s × 10⁻⁵
10	1.3060	2.7276	1.3065	1.214
20	1.0016	2.0919	1.0035	0.933
30	0.7972	1.6650	0.8007	0.744
40	0.6527	1.3632	0.6579	0.611
50	0.5465	1.1414	0.5531	0.514
60	0.4660	0.9733	0.4740	0.441
70	0.4035	0.8428	0.4127	0.384
80	0.3540	0.7394	0.3643	0.339
90	0.3142	0.6562	0.3255	0.303
100	0.2816	0.5881	0.2938	0.273

Table 1.2 shows the dynamic and kinematic viscosity of water at various temperatures. At 20°C, the kinematic viscosity of water is:

$$\upsilon = 10^{-6} \, m^2 \, s$$

Or

$$\upsilon = 10^{-5} \, ft^2 \, s$$

1.5 FRICTION LOSS

Julius Weisbach (1806-1871),   Henry Darcy (1803-1858)

In solving hydraulic problems or designing hydraulic systems, one may use one, two or all three principles of hydraulics depending on the nature of the problem/ system. When conservation of energy (Equation (1.1), or (1.2)) is used, it is necessary to calculate head loss (h_L). The head loss consists of friction loss and minor loss (h_f & h_m). The Darcy–Weisbach equation, named for Julius Weisbach (1806– 1871) and Henry Darcy (1803–1858), is used to calculate friction loss in pipes as follows:

$$h_f = f\left(\frac{L}{D}\right)\frac{V^2}{2g} \tag{1.9}$$

where h_f is friction loss, D is the inside diameter of the pipe, L is the length of the pipe, V is the average velocity of the flow stream and f is calculated from the Moody diagram shown below.

If the flow is laminar, the f value can be calculated as:

$$f = \frac{64}{Re} \tag{1.10}$$

where Re is the Reynolds number

Example 1.2

A 40-mm diameter circular pipe of length 10 m carries water at 20°C and velocity of 5 cm/s. The pipe outlet is free flow.

A. Calculate the Reynolds number
B. Is the flow laminar or turbulent?
C. What is the pressure at the end of the pipe?
D. What is the flow rate?

Solution:
The kinematic viscosity of water at 20°C is $v = 1 \times 10^{-6}$ m²/sec. Re = 2000, $D = 0.04$ m.

$$Re = \frac{DV}{v} = \frac{0.04(0.05)}{1 \times 10^{-6}} = 2000$$

The flow is laminar.
The pressure at the outlet is zero.
The flow is: $Q = A.V$.
The flow rate is:

$$Q = AV = \frac{\pi}{4}(0.04)2 \times 0.05\,\text{m}/\sec = 6.28 \times 10^{-5}\,\text{m}^3/\sec$$

Example 1.3

Water flows through a pipe of diameter 0.04 m. The Reynolds number for the flow is 1000. The pipe is connected to a tank with water level elevation of 2 m. The pipe is 10 m long and is flowing free at the outlet at elevation 0.0.

A. Determine the flow rate
B. What would be the pressure half way along the pipe?

Solution:

Using Equation (1.2),

$$2+0+0 = 0+0+\frac{v^2}{2g}+\left(\frac{64}{1000}\times\frac{10}{0.04}\times\frac{v^2}{2g}\right), \text{ and from here:}$$

$v = 1.52$ m/s, $Q = V.A. = 1.52 \times \frac{\neq}{4}(0.04)^2 = 0.0019$ m³/sec

To calculate the pressure at the halfway point, use Equation (1.2):

$$2+0+0 = \frac{p}{\gamma}+\frac{1.52^2}{2g}+0+\left(\frac{64}{1000}\times\frac{5}{0.04}\times\frac{1.52^2}{2g}\right)$$

and $\frac{p}{\gamma} = 0.94$ m

What is the pressure in PSI?

Frictional energy loss in pipelines is commonly considered the major loss. h_f is the loss of head due to pipe friction and viscous loss in the flowing water.

The Moody diagram is used to calculate the friction factor, f, in the Darcy–Weisbach equation (Equation (1.9)). The friction factor is then used to calculate friction loss. In the Moody diagram, the friction factor f is calculated by using relative roughness (e/D) and the Reynolds number Re.

Relative roughness is the ratio of pipe roughness e to pipe inside diameter D. Table 1.3 shows values of roughness, e, for various pipe materials.

TABLE 1.3
Roughness Height, e, for Various Pipe Materials

Pipe Material	e (mm)	e (ft)
Glass, brass, copper	0.0015	0.000005
Commercial steel (coated)	0.0048	0.000016
Commercial steel (new)	0.045	0.00015
Wrought Iron	0.045	0.00015
Asphalt Cast Iron (new)	0.12	0.004

(Continued)

TABLE 1.3   (CONTINUED)

Pipe Material	e (mm)	e (ft)
Galvanized iron	0.15	0.0005
Cast iron (new)	0.26	0.00085
Wood Stave (new)	0.18 to 0.9	0.0006 to 0.003
Concrete (smooth)	0.18	0.0006
Riveted steel	0.90	0.003
Corrugated Metal	45	0.15
PVC/HDPE	0.0015	0.000005

Example 1.4

Compute the discharge capacity of a 3-m diameter, wood-stave pipe in its best condition (e = 0.3 mm), carrying water from a tank to an open discharge delivery point as shown. The head at the tank is 2 m and the length of the pipe is 1 km. The flow is free discharge at the end.

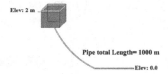

Elev: 2 m

Pipe total Length= 1000 m

Elev: 0.0

Solution:

From Equation (1.9), the friction head loss in the pipe is

$$h_f = f\left(\frac{L}{D}\right)\frac{V^2}{2g}$$

Hence:
From Table 1.1, taking e = 0.3 mm, we obtain for the 3-m, wood-stave pipe:

$$\frac{e}{D} = 0.0001$$

Using kinematic viscosity of water $\upsilon = 10^{-6}$ m²/sec. Therefore,

$$Re = \frac{DV}{\upsilon} = \frac{3V}{10^{-6}} = 3\times10^6 V \text{ (Equation 1.8 above)}$$

Using the energy equation and ignoring minor losses:

$$E_1 = E_2 + h_f$$

$$2 = 0 + \frac{V^2}{2(9.81)} + f\left(\frac{1000}{3}\right)\frac{V^2}{2(9.81)}$$

$$\text{Function} = 2 - \frac{V^2}{2(9.81)}\left(f\frac{1000}{3} + 1\right) = 0$$

The non-linear function described above can be solved using three possible options: the iterative method; trial and error method; and numerical optimization method.

The iterative method is used here by re-arranging the function as

$$V^2 = \frac{2(2)(9.81)}{\left(1 + f\frac{1000}{3}\right)}$$

The non-linear equation shown above can then be solved iteratively as shown in the following table.

V_1 (m/s)	Re	f (from Moody diagram)	V_2 (m/s)
1 (assumed)	3×10^6	0.013	2.71
2.71	8.13×10^6	0.0122	2.78
2.78	8.35×10^6	0.0121	2.78

$V = 2.78$ m/s, which is considered close enough to the previous assumed value and will be used to calculate the flow.

This value and the cross-sectional area give the discharge.

$$Q = AV = \pi\left(\frac{3^2}{4}\right) \times (2.78) = 19.64\,\text{m}^3/\text{sec}$$

Another way to solve the non-linear function shown above is by trial and error which results in the function approaching zero.

The following example shows how to use trial and error to solve a problem.

Example 1.5

Compute the required diameter for a wood-stave pipe in its best condition ($e = 0.3$ mm) to carry 22 m³/s of water at 10°C. The water is to be carried from an elevation of 3 m to elevation of 1 m. The length of the pipe is 1 km.

Solution:

Using the energy equation (Equation (1.2)) and ignoring minor losses:

$$E_1 = E_2 + h_f$$

$$0 + 0 + 3 = 0 + \frac{V^2}{2g} + 1 + f\left(\frac{1000}{D}\right)\frac{V^2}{2(9.81)}$$

$$\text{Function} = 2 - \frac{V^2}{2g}\left(1 + f\ \frac{1000}{D}\right) = 0.0$$

and

$$\text{Re} = \frac{DV}{\upsilon} = \frac{D.V}{1.31 \times 10^{-6}}$$

D	V	Re	f	Function
2	7.01	1.07×10^7	0.0121	−15.66
3	3	6.87×10^6	0.0121	−0.31
3.1	2.92	6.91×10^6	0.0121	−0.13
3.2	2.74	6.7×10^6	0.0122	0.16

Therefore, a diameter of 3.2 m can deliver 22 m³/s as desired.

1.6 MATHEMATICAL EQUATION FOR MOODY DIAGRAM

Alternatively, the graphical Moody diagram can be replaced by mathematical equations which can be used to solve hydraulic problems using computer programs. The procedure is described below.

1.6.1 DARCY'S FRICTION LOSS EQUATION

The Darcy–Weisbach equation relates friction loss in a conduit (h_f), to the velocity head $\left(\frac{V^2}{2g}\right)$ and to the ratio of pipe length (L) to pipe diameter (D), accounting for fluid properties, conduit properties and type of flow regime with the friction factor (f). If water flows from point 1 to point 2 through a pipe with diameter D, then the following equation describes the process:

$$dE = E_1 - E_2 = \left(f\frac{L}{D} + \Sigma K\right)\frac{V^2}{2g} \tag{1.11}$$

where

$E_1 - E_2$ is the difference in energy between two points in meter or (feet)
f is the dimensionless friction factor
L is the pipe length in meters or (feet)
D is the conduit diameter in meters or (feet)
ΣK is sum of dimensionless minor loss coefficients
V is the velocity in m/s (ft/s)
g is the acceleration of gravity in m/s² (ft/s²).

Depending on the flow regime the friction factor is a function of the Reynolds number for laminar flow, and both the Reynolds number and the relative roughness of the pipe for turbulent flow. The Reynolds number is defined as the ratio of the inertia force of an element of fluid to the viscous force. This dimensionless number depends on the fluid density, fluid viscosity, pipe diameter and average flow velocity. The expression for the Reynolds number, denoted by Re in this chapter, is as follows:

$$Re = \frac{\rho VD}{\mu} \tag{1.12}$$

$$Re = \frac{4Q}{\pi D\upsilon} \tag{1.13}$$

where
Re is the Reynolds number (dimensionless)
ρ is the fluid density in kg/m^3 (slugs/ft^3)
μ is the fluid dynamic viscosity in N s/m^2 (lb-s/ft^2)
υ is the kinematic viscosity (μ/ρ) in m^2/s (ft^2/s)
V is the flow velocity in m/s (ft/s)
Q is the flow rate in m^3/s (ft^3/s)
D is the pipe diameter in meters (feet).

The relative roughness is given by the ratio of the average wall roughness (e) to the pipe diameter (D). Average wall roughness values for new commercial pipes have been widely published in the literature and will not be included here.

Probably the most widely used method of evaluating the friction factor employs the Moody diagram, which relates the friction factor (f) to the Reynolds number (Re) and the relative roughness (e/D). This diagram was developed experimentally y L. F. Moody in 1944. Computerized calculation of the friction factor, however, requires equations for different flow regimes. For laminar flow conditions, where the Reynolds number is below 2000, the friction factor (f) is given by the equation:

$$f = \frac{64}{Re} \tag{1.14}$$

where
f is the dimensionless friction factor,
Re is the dimensionless Reynolds number.

At Reynolds numbers between 2000 and 4000 the flow is in a critical range and the friction factor cannot be accurately predicted. Reynolds numbers above 4000 indicate a region of turbulent flow. The turbulent zone is further divided into a 'smooth pipe' boundary, a transition region and a region of complete turbulence.

Along the smooth pipe boundary the friction factor is a function of the Reynolds number only and in the fully turbulent region, the friction factor is a function of the relative roughness. In the transition zone, however, the friction factor is a function of both the Reynolds number and the relative roughness.

C. F. Colebrook developed a relation for the friction factor in the transition zone. This equation, which is of implicit nature, includes two terms inside the log function to account for both the smooth pipe boundary and the fully turbulent flow region. The implicit form of Colebrook's equation can be written as follows:

$$\frac{1}{\sqrt{f}} = 1.14 - 2\log\left(\frac{e}{D} + \frac{9.35}{\text{Re}\sqrt{f}}\right) \tag{1.15}$$

where
  f is the dimensionless friction factor
  Re is the dimensionless Reynolds number
  D is the pipe diameter in meters (feet)
  e is the wall roughness in meters (feet).

Equation (1.15) above accounts for fully turbulent flow as the second term inside the parenthesis becomes very small. Also, very small values of e/D reduce the equation to that of the smooth pipe boundary.

Explicit equations have been developed to compute the friction factor as an alternative to Colebrook's equation. P. K. Swamee and A. K. Jain developed the following equation:

$$f = \frac{0.25}{\left[\log\left(\dfrac{1}{3.7\left(\dfrac{D}{e}\right)} + \dfrac{5.74}{\text{Re}^{0.90}}\right)\right]^2} \tag{1.16a}$$

or

$$f = \frac{0.25}{\left[\log\left(\dfrac{e}{3.7D} + \dfrac{5.74}{\text{Re}^{0.90}}\right)\right]^2} \tag{1.16b}$$

where

  F is the dimensionless friction factor
  Re is the dimensionless Reynolds number
  D is the pipe diameter in meters (feet)
  e is the wall roughness in meters (feet)
  D/e is greater than 100 and less than 167,000
  Re is greater than 4000 and less than 10^8.

The solution of the friction loss h_f in Darcy's Equation is explicit. If the unknown is the flow rate, however, an iterative process is required because the Reynolds number cannot be explicitly computed since it is a function of velocity and therefore flow rate. This problem is further complicated when the diameter is the unknown in Darcy's Equation, since both the Reynolds number and the relative roughness cannot be calculated. Regardless of the unknown being considered, a technique is available to the design engineer to compute any of the three scenarios just described by treating Darcy's Equation as an implicit function as follows:

$$f(x) = \left(f\frac{L}{D} + \Sigma K \right)\frac{V^2}{2g} - dE = 0 \tag{1.17}$$

$$V = \frac{4Q}{\pi D^2} \tag{1.18}$$

where
    $f(x)$ is an implicit function for Darcy's equation
    x is a dummy variable representing dE, Q or D
    dE is the energy difference between the two points meters (feet)
    L is the pipe length in meters (feet)
    D is the conduit diameter in meters (feet)
    ΣK is a dimensionless minor loss coefficient
    V is the flow velocity in m/s (ft/s)
    Q is the flow rate in m³/s (ft³/s)
    g is the acceleration of gravity in m/s² (ft/s²)

The non-linear function (1.17) can be solved iteratively for Q, D, V or "f" without the need to refer to Moody's diagram.

1.7 EMPIRICAL CLOSED-FORM SOLUTION OF ENERGY CONSERVATION EQUATION (EQUATION (1.2))

1.7.1 HAZEN–WILLIAM EQUATION

The Hazen–William equation is a closed form solution of the energy conservation equation which is valid only for water, as opposed to the Darcy–Wiesbach equation which is valid for any fluid. The H–W equation is also limited to:

$D > 2$ inches
$V < 3$ m/s

Three forms of the H–W equation using metric, English and conventional units are given below. In English units:

$$V = 1.318C_{HW}R_h^{0.63}S^{0.54} \tag{1.19}$$

where C_{HW} is the Hazen–William coefficient (Table 1.4), S is the slope of the energy gradient line (EGL), or head loss per unit length of the flow in the pipe ($S = h_f/L$), and R_h is the *hydraulic radius*, defined as the water cross-sectional area, A, divided by the *wetted perimeter*, P. For a circular pipe, $A = \pi D^2/4$ and $P = \pi D$, the hydraulic radius is,

$$R_h = \frac{A}{P} = \frac{\frac{\pi D^2}{4}}{\pi D} = \frac{D}{4} \tag{1.20}$$

The Hazen–Williams coefficient, C_{HW}, is not a function of the flow conditions (i.e., Reynolds number). Its values range from 150 for very smooth, straight pipes down to 60 for unlined, old pipes. The values of C_{HW} for commonly used water carrying conduits are listed in Table 1.4.

TABLE 1.4
Hazen–William Coefficients

Material	C_{HW}
Aluminum	130–150
Asphalt lining	130–140
Brass	130–140
Brick	90–100
Cast iron, new unlined	130
Cast iron, old	64–83
Cast iron, asphalt coated	100
Cast iron, cement lined	140
Cast iron, bituminous lined	140
Cast iron, sea-coated	120
Cast iron, wrought plain	100
Concrete	100–140
Concrete lined, steel forms	140
Concrete lined, wooden forms	120
Concrete, old	100–110
Copper	130–140
Corrugated metal	60
Ductile iron pipe (DIP)	140
Ductile iron, cement lined	120
Galvanized iron	120
Lead	130–140
Metal pipes, very to extremely smooth	130–140
Plastic	130–150
Polyethylene, PE	140
Polyvinyl chloride, PVC	150
Smooth pipes	140
Steel new unlined	140–150

TABLE 1.4   (CONTINUED)

Material	C_{HW}
Steel, corrugated	60
Steel, welded and seamless	100
Steel, interior riveted, no projecting rivets	110
Steel, vitrified, spiral-riveted	90–110
Steel, welded and seamless	100
Wrought iron, plain	100
Wood stave	110–120

In SI (metric) units, the Hazen–Williams equation is in the following form:

$$V = 0.85 C_{HW} R_h^{0.63} S^{0.54} \tag{1.21}$$

In conventional units the H–W equation is in the following form:

$$h_f = 10.5(L).\left(\frac{Q}{C}\right)^{1.852}.D^{-4.87} \tag{1.22}$$

and

$$Q = 0.28.C.D^{2.63}.S^{0.54} \tag{1.23}$$

where

h_f is the friction loss, ft
L = pipe length, ft
C or C_{HW} = H–W coefficient (Table 1.4)
Q = flow rate, gpm
D = diameter, inches
S = slope of energy grade line (h_f/L)

Example 1.6

A 300-ft-long pipe with D = 12 inches and C_{HW} = 120 is connected to a reservoir. The water level elevation in the reservoir is 30 ft above the delivery point. The flow is free discharge.

Determine the flow rate.

Solution:

Using Equation (1.23), and ignoring velocity component,

$$Q = 0.28(120).12^{2.63}\left(\frac{30}{300}\right)^{0.54} = 6,677\,\text{gpm}$$

1.7.2 MANNING EQUATION

The Manning equation is normally used for open channels, but it can also be used in pipe flow. It is common to use the Manning equation for flow through culverts or drains. Culverts are pipes connecting sections of open channel together. This will be discussed further in Chapter 5.

The Manning equation was originally developed in metric units. and has been extensively used for open channel designs. The Manning equation in the metric (SI) system is:

$$V = \frac{1}{n} R_h^{2/3} S^{1/2} \qquad (1.24)$$

where n is Manning's coefficient of roughness, specifically known to hydraulic engineers as Manning's n. (Table 1.5). In English units, the Manning equation is:

$$V = \frac{1.49}{n} R_h^{2/3} S^{1/2} \qquad (1.25)$$

where the hydraulic radius, R_h, is measured in feet, and the velocity is measured in ft/s. Table 1.5 shows values for Manning roughness which are independent of units.

TABLE 1.5
Typical Manning Roughness Coefficient, *n*, for Various Materials

Surface Material	Manning Roughness Coefficient
Asphalt	0.016
Brickwork	0.015
Cast iron, new	0.012
Clay tile	0.014
Concrete, steel forms	0.011
Concrete (cement), finished	0.012
Concrete, wooden forms	0.015
Concrete, centrifugally spun	0.013
Earth channel	0.025
Earth channel, clean	0.022
Earth channel, gravelly	0.025
Earth channel, weedy	0.030
Earth channel, stony, cobbles	0.035
Floodplains, light brush	0.050
Floodplains, heavy brush	0.075

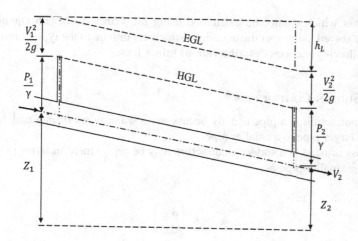

FIGURE 1.2   EGL and HGL along a pipe.

1.7.3   ENERGY GRADE LINE (EGL) AND HYDRAULIC GRADE LINE (HGL)

The Energy Grade Line (EGL) represents energy available along the pipe. It is used to show the change in energy. The EGL at any point along the stream is:

$$E = \frac{V^2}{2g} + \frac{P}{\gamma} + z_1 \qquad (1.26)$$

The Hydraulic Grade Line (HGL) represents the EGL minus the kinetic energy as:

$$E = \frac{P}{\gamma} + z_1 \qquad (1.27)$$

The Hydraulic Grade Line (HGL) is used to demonstrate the pressure condition in the pipe. If the HGL falls below the pipe, it shows that the pipe is under negative pressure, a situation which needs to be avoided in pressurized systems.

Figure 1.2 shows the EGL and HGL along a pipe.

1.8   CALCULATING MINOR LOSSES

Minor losses occur whenever the flow changes direction. As long as flow is moving along a straight line, the head loss can be calculated using Darcy–Weisbach equation, but once there is a change in direction of flow such as contraction (water moving from a large pipe to small pipe), expansion (water moving from small pipe to large pipe), or through a bend in pipe or through a valve, then one needs to calculate the minor loss and add it to the overall head losses. In reality, minor losses are actually friction loss which cannot be calculated using the Darcy–Weisbach equation and need to be solved using specific equations which are unique to the type of minor loss.

The following sections describe various minor losses.

1.8.1 SUDDEN CONTRACTION

Sudden contraction in a pipe usually occurs when water flows from a tank to a pipe or from a larger pipe to a smaller pipe.

The loss of head in a sudden contraction may be represented in terms of velocity head in the smaller pipe as:

$$h_m = K_c \left(\frac{V_2^2}{2g} \right) \tag{1.28}$$

where

V_2 is the velocity in the second/smaller pipe.

The value of minor loss coefficient (K_c) in Equation (1.28) can be calculated using Table 1.6.

1.8.2 GRADUAL CONTRACTION

Head loss due to pipe contraction may be greatly reduced by introducing a gradual pipe contraction, shown in Figure 1.3. The head loss in this case may be expressed as

$$h_c' = K_c' \left(\frac{V_2^2}{2g} \right) \tag{1.29}$$

The values of K_c' vary with the transition angle, α, and the area ratio A_2/A_1, as shown in Figure 1.4.

TABLE 1.6

Values of Coefficient K_c for Sudden Contraction

Velocity in Smaller Pipe V (m/sec)	Ratio of Smaller to Larger Pipe Diameters, D_2/D_1				
	0.0	0.2	0.4	0.6	0.8
1	0.49	0.48	0.42	0.28	0.11
2	0.48	0.47	0.41	0.28	0.10
3	0.47	0.45	0.40	0.28	0.09
6	0.45	0.43	0.38	0.27	0.07

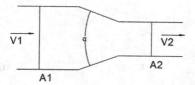

FIGURE 1.3 Pipe gradual contraction.

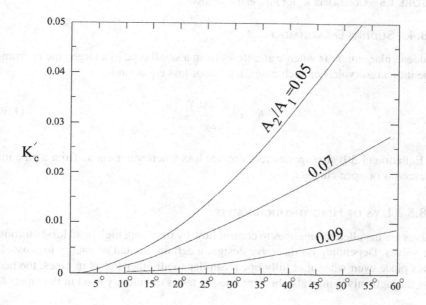

FIGURE 1.4 Coefficient K_c' for pipe confusors.

1.8.3 ENTRANCE LOSSES

The loss of head at the entrance to a pipe from a large reservoir is a special case of loss of head due to contraction. Since the water cross-sectional area in the reservoir is very large compared to that of the pipe, a ratio of contraction of zero may be taken. For a square-cornered entrance, where the entrance of the pipe is flush with the reservoir wall; as shown in Figure 1.5(a), the K_c values shown for $D_2/D_1 = 0.0$ in Table 1.6 are used. The approximate values for the coefficient of entrance loss K_e for different conditions are shown in Figure 1.5a–c.

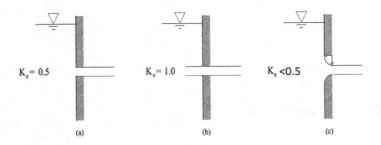

FIGURE 1.5   Coefficient K_e for Pipe entrance loss.

1.8.4   SUDDEN ENLARGEMENT

Sudden enlargement is when water flows from a small pipe to a large pipe or from a pipe into a reservoir. For such cases, the minor loss equation is:

$$h_m = \frac{(V_1 - V_2)^2}{2g} \tag{1.30}$$

Equation (1.30) also applies for entrance loss when water enters from a pipe into a reservoir or open channel.

1.8.5   LOSS OF HEAD THROUGH VALVES

Valves are installed in pipelines to control flow by imposing high head losses through the valves. Depending on the valve design, a certain amount of energy loss usually takes place even when it is fully open. Similar to all other losses in pipes, the head loss through valves may also be expressed in terms of velocity head in the pipe:

$$h_v = k_v \frac{V^2}{2g} \tag{1.31}$$

The values of k_v, vary with the design of the valves. When designing a hydraulic system, it is necessary to determine the required loss of head through each valve in the system. The values of k_v, for common valves are shown in Figure 1.6.

1.8.6   MINOR LOSS THROUGH BENDS

Minor loss through a bend depends on its type. Figure 1.7 shows minor loss coefficients for some common bends. For special conditions, the minor loss coefficient can be empirically measured through laboratory experiments.

The following example shows how a hydraulic problem comprising of friction loss and minor losses can be solved.

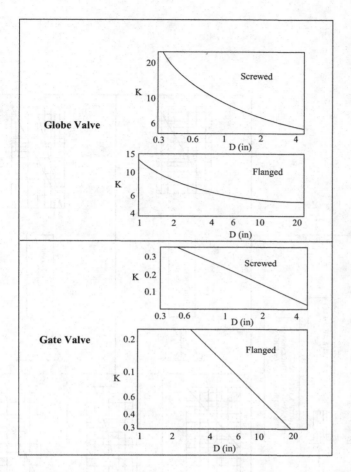

FIGURE 1.6 Minor loss coefficients in valves.

Example 1.7

Figure 1.8 shows two sections of cast-iron pipe, connected in series, that bring water from a reservoir and discharge it into open air at a location 100 m below the water surface elevation in the reservoir through a valve with minor loss coefficient of "$k_v = 10$". If the water temperature is 10°C, and square connections are used, determine the discharge.

Solution:

A conservation of energy equation can be written for Section 1 at the reservoir and Section 3 at the discharge end

$$\frac{V_1^2}{2g} + \frac{P_1}{\gamma} + z1 = \frac{V_2^2}{2g} + \frac{P_3}{\gamma} + z_3 + h_L$$

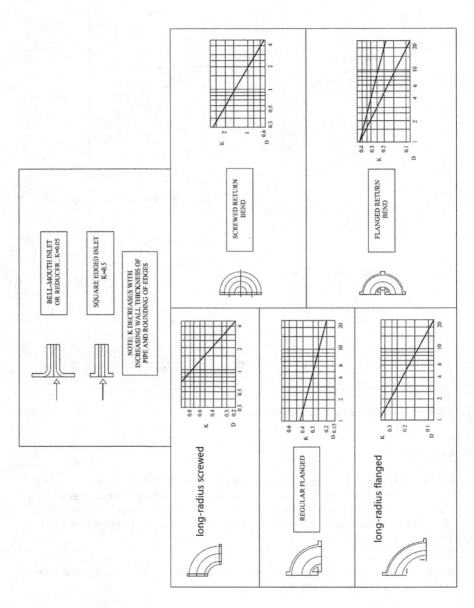

FIGURE 1.7   Minor loss coefficients for bends

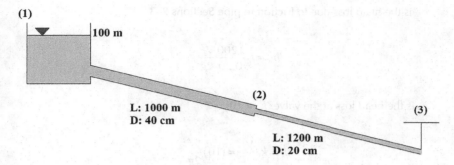

FIGURE 1.8 Flow through variable pipe sizes.

Selecting the reference datum at Section 3, $z_3 = 0$. Since the reservoir and the discharge end are both exposed to atmospheric pressure and the velocity head the reservoir can be neglected, we have,

$$E_1 = 100 = \frac{V_2^2}{2g} + h_L$$

The total available energy, 100 m of water column, is equal to the velocity head at the discharge end plus all the head losses incurred in the pipeline system. This relationship may be expressed as:

$$h_e + h_{f_1} + h_c + h_{f_2} + h_v + \frac{V_2^2}{2g} = 100$$

where h_e is the head loss at the entrance. For a square-cornered entrance, Figure 1.4 gives:

$$h_e = (0.5)\frac{V_1^2}{2g}$$

h_f is the head loss due to friction in pipe Sections 1–2. From Equation (1.16a),

$$h_{f_1} = f_1\frac{1000}{0.40}\frac{V_1^2}{2g}$$

h_c is the head loss due to sudden contraction at Section 2. From Table 1.6 assume $K = 0.33$ for a first trial).

$$h_c = k_c\frac{V_2^2}{2g} = 0.33\frac{V_2^2}{2g}$$

h_{f_2} is the head loss due to friction in pipe Sections 2–3

$$h_{f_2} = f_2 \frac{1200}{0.20} \frac{V_2^2}{2g}$$

h_v is the head loss at the valve, $k_v = 10$:

$$h_v = k_v \frac{V_2^2}{2g} = (10)\frac{V_2^2}{2g}$$

Therefore, a function can be developed as:

$$\text{Function} = 100 - \left[\left(1+10+f_2\frac{1200}{0.20}+0.33\right)\frac{V_2^2}{2g} + \left(f_1\frac{1000}{0.40}+0.5\right)\frac{V_1^2}{2g}\right] = 0.0$$

From the continuity equation, we have

$$A_1V_1 = A_2V_2$$

$$\frac{\pi}{4}(0.4)^2 V_1 = \frac{\pi}{4}(0.2)^2 V_2$$

The above relations give

$$\text{Function} = 100 - \left[\left(1+10+f_2\frac{1200}{0.20}+0.33\right)\frac{V_2^2}{2g} + \left(f_1\frac{1000}{0.40}+0.5\right)\times 0.063\frac{V_2^2}{2g}\right] = 0.0$$

$$V_2^2 = \frac{1962}{11.36+156.25f_1+6000f_2}$$

where $v = 1.31 \times 10^{-6}$ at 10°C. For the 40-cm pipe, $e/D = 0.00065$; $e/D = 0.0013$ for the 20-cm pipe.
To evaluate f_1 and f_2, we use the e/D and Re values shown below:

$$Re_1 = \frac{D_1V_1}{v} = \frac{0.4}{1.31\times10^{-6}}V_1 = \frac{0.4}{1.31\times10^{-6}}(0.25V_2) = 7.63\times10^4 V_2$$

$$Re_2 = \frac{D_2V_2}{v} = \frac{0.20}{1.31\times10^{-6}}V_2 = 1.53\times10^5 V_2$$

Using the iteration method, and assuming a $V_2 = 4$ m/s:

V_2	f_1	f_2	V_2
4	0.0178	0.0205	3.78
3.78	0.019	0.021	3.74

$V_2 = 3.78$ m/s
$V_1 = 0.25(V_2) = 0.25(3.78) = 0.94$ m/s

Hence,

$$Re_1 = 7.63 \times 104(3.78) = 2.88 \times 105; f1 = 0.019$$

$$Re_2 = 1.53 \times 105(3.78) = 5.78 \times 105; f2 = 0.021$$

And since the difference between initial and final V_2 is 1%, we can accept $V_2 = 3.74$ m/s as the answer and the discharge is calculated as:

$$Q = A_2V_2 = \frac{\pi}{4}(0.2)2(3.74) = 0.12 \, m^3 / sec$$

In solving this problem, we initially assumed a minor loss coefficient, K, of 0.33. Using the final velocity values the K value from Table 1.6, can be interpolated as 0.33. Otherwise, we would need to assume another K value and repeat the iteration.

1.9 DESIGN PRINCIPLES

In this chapter, we have discussed the concept of pipe design mainly based on gravitational flow. The design of a structure including hydraulic structure is based on three principles, described below:

1. Design should be functional
2. Design should be economical
3. Design should be ethical

The concept of functionality of the design implies that the design of a structure should meet the objectives of the design. For example, if we want to design a pipeline to carry a certain amount of flow from a source to a destination, the pipe dimensions including diameter and length should be determined such that the objective is met.

The concept of economical design implies that while the design delivers the objectives, it should also be as economical as possible. The cost of a hydraulic structure depends on three factors:

$$Cost = Capital \; cost + Energy \; cost + Maintenance \; cost$$

If a hydraulic structure functions based on gravitational or hydraulic pressure without the need for additional energy, the pipe dimensions (e.g., length, diameter, material) should be selected such that the capital cost and maintenance cost is minimal. If energy is required to operate the system, then an annual cost can be calculated and minimized based on the following equation.

$$\text{Annual cost} = \text{CRF} * (\text{Capital cost}) + \text{Energy cost} + \text{Maintenance cost} \quad (1.32)$$

in which CRF is the capital recovery factor which is used to amortize the initial capital cost into yearly cost, based on the interest rate and the life expectancy of the structure. This concept will be described further in Chapter 3.

The ethical principles of design cover a wide range of issues which may depend on local factors, including

a. Safety
b. Standards
c. Legal and environmental issues
d. Sustainability
e. Aesthetic

Safety is a major issue in the design and maintenance of structures. For example, it is not sufficient to just determine the necessary diameter of a pipe. It is also necessary to ensure that the pipe has sufficient wall thickness and tensile strength to withstand both normal and transient pressures in the system. This will be discussed in detail in Chapter 2.

1.10 SIMPLIFIED PIPE ECONOMICS

The principles of design require us to design structures such that they are functional, economical and safe (ethical). While the economics of design involves detailed analysis, a simple empirical concept of pipe design is shown in Figure 1.9. The economic size of a pipe is a function of initial pipe cost amortized over the life expectancy of the pipe and the energy cost associated with running the water through the pipe. Figure 1.8 illustrates the relationship between the initial cost, the energy cost and the optimum pipe size, similar to the concept of supply and demand in economics.

As shown in Figure 1.8, there is an optimum point at which the combined annual cost of material and energy is minimized. Economic analysis in the past has shown that this minimum cost is achieved when the velocity of water is maintained around 7–10 ft/s (2–3 m/s), although it is expected that the optimum point will change due to changes in cost of material and energy. Experience has shown that due to the dependency of material cost on energy cost, the optimum velocity does not change much due to continuous dependence of material cost on energy cost. However, in practice, we often deal with complex multicomponent systems which require a more detailed economic analysis. This issue will be discussed further in Chapter 3 on pump design.

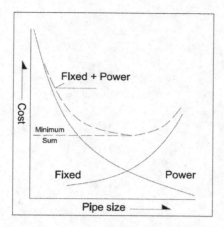

FIGURE 1.9   Influence of pipe size on fixed and power costs associated with cost-effective pipe size selection.

REFERENCES

1. Colebrook, C. F. (1939). Turbulent Flow in Pipes, with Particular Reference to the Transition Region Between the Smooth and Rough Pipe Laws. *Journal of the Institution of Civil Engineers* 11(4): 133–156. doi:10.1680/ijoti.1939.13150. ISSN: 0368-2455.
2. Swamee, P. K. and Jain, A. K. (1976). Explicit Equations for Pipe-Flow Problems. *Journal of the Hydraulics Division* 102(5): 657–664. doi:10.1061/JYCEAJ.0004542.
3. Moody, L. F. (1944). Friction Factors for Pipe Flow. *Transactions of the ASME* 66(8): 671–676.

2 Design Safety

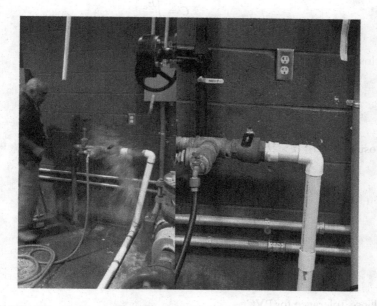

FIGURE 2.1   Water hammer demonstration, NMSU Hydraulics Lab. "Pipe collapsing due to negative water hammer caused by sudden closure".

2.1   SUMMARY EQUATIONS

Pressure Tolerance

$$P = \frac{\text{HDB}}{\text{SDR} - 1}$$

HDB = Hydrostatic Design Basis or Tensile Strength (4000 for PVC, 35,000 for mild steel, for other material refer to Table 2.1)

$$\text{SDR} = \frac{\text{OD}}{\text{th}}$$

DOI: 10.1201/9781003287537-2

$$\frac{1}{E_c} = \frac{1}{E_b} + \frac{D}{E_p.th}$$

$$C = \sqrt{\frac{E_c}{\rho}}$$

t = closure time, if ($t \le 2L/C$)

$$\Delta P = \frac{E_c}{AL}\left[V_0 A\left(\frac{L}{C}\right)\right] = \frac{E_c V_0}{C} = \rho.C.dV = \rho.C.V$$

If closure time is: ($t > 2L/C$), then

$$N = \left(\frac{\rho L V_0}{P_0 t}\right)^2$$

$$\Delta P = P_0\left[\frac{N}{2} + \sqrt{\frac{N^2}{4} + N}\right]$$

Collapse Tolerance for PVC

$$P_0 = \left(\frac{(2E)0.75}{(1-u^2)\left[\frac{D_0}{t}-1\right]^3}\right)$$

Point Load

$$\sigma_z = \frac{3Q \times 1}{2\pi Z^2\left[1+\left(\frac{r}{z}\right)^2\right]^{\frac{5}{2}}} = \frac{Q}{Z^2} \times I_B$$

2.2 DESIGN CRITERIA

1. Design should be functional (meet the objectives).
2. Design should be economical.
3. Design should be ethical (safe, meet legal, aesthetic and sustainability standards).

2.2.1 Pipe Safety

Pipes may fail due to the following:

1. Excessive internal hydrostatic pressure
2. Water hammer ±
3. Excessive external load
4. Fatigue
5. Freezing

Pipeline design should ensure that the system can handle the following scenarios:

1. The system should tolerate the maximum hydrostatic pressure in the system.
2. The system should tolerate transient surges that may occur due to positive or negative water hammer resulting from rapid change in velocity (dV/dt).
3. The system should tolerate excessive external load.
4. The system should be protected against damage due to freezing.

2.3 WATER HAMMER PHENOMENON IN PIPELINES

Sudden change of flow or velocity in a pipe due to valve closure, valve opening, pipe breakage, pump turn-on or turn-off can result in excessive change in pressure which can cause a pipe to burst or collapse. Excessive change in pressure will also cause pipe fatigue which can lead to its eventual failure. Excessive pressure change may also fracture the pipe walls or cause other damage to the pipeline system. The possibility of such pressure occurring, its magnitude, and the propagation of the pressure wave form must be carefully estimated in pipeline design.

The sudden change of pressure can be demonstrated by Newton's second law of motion. When the flow mass changes its velocity at the rate dV/dt, according to Newton's second law of motion,

$$F = m \frac{dV}{dt} \tag{2.1}$$

If the velocity of the entire water column could be reduced to zero instantly, Equation (2.1) would become

$$F = \frac{m(V_0 - 0)}{0} = \frac{mV_0}{0} = \infty \tag{2.2}$$

The resulting force (hence, pressure) would be infinite. Fortunately, such an instantaneous change is almost impossible because a mechanical valve requires a certain amount of time to complete a closure operation. In addition, neither the pipe walls nor the water column involved are perfectly rigid under high pressure. The elasticity of both the pipe walls and the water column plays a very important role in the water hammer phenomenon. In fluid mechanics, Newton's second law of motion can also be rewritten as:

$$\Sigma F = \rho \times Q \times dV \qquad (2.3)$$

in which ρ is the density of fluid, Q is the flow rate, and dV is change in velocity. This is what is called the impulse momentum principle.

Water hammer, either positive or negative, can result in three potential problems: it can burst the pipe; it can collapse the pipe; or it can fatigue the pipe and eventually break it.

Consider a pipe length L with inside diameter D, wall thickness, th, and modulus of elasticity E_p. Immediately following the valve closure, the water in close proximity to the valve is brought to rest. The sudden change of velocity in the water mass causes a local pressure increase. As a result of this pressure increase, the water columns in this section are somewhat compressed, and the pipe walls expand slightly due to the corresponding increase of stress in the walls.

Both of these phenomena help provide a little extra volume, allowing water to enter the section continuously until it comes to a complete stop. The next section immediately upstream is involved in the same procedure an instant later. In this manner, a wave of increased pressure propagates up the pipe toward the reservoir, as shown in Figure 2.1. When this pressure wave reaches the upstream reservoir, the entire pipe is expanded and the water column within is compressed by the increased pressure. At this very instant the entire water column within the pipe comes to a complete halt.

Obviously, this transient state cannot be maintained because the pressure in the pipe is much higher than the pressure in the open reservoir. The halted water in the pipe begins to flow back into the reservoir as soon as the pressure wave reaches the reservoir. This process starts at the reservoir end of the pipe and a decreased pressure wave travels downstream toward the valve. During this period, the water behind the wave front moves in the upstream direction as the pipe continuously contracts and the columns decompress. The time required for the wave to return the valve is $2L/C$, where C is the speed of wave travel along the pipe. It is also known as *celerity*.

The speed of pressure wave travel in a pipe depends on the composite modulus of elasticity of water E_b, and the modulus of elasticity of the pipe wall material E_p. The relationship may be expressed as

$$C = \sqrt{\frac{E_c}{\rho}} \qquad (2.4)$$

where E_c is the composite modulus of elasticity of the water-pipe system and ρ is the density of water (Modulus of elasticity of water is 2.2×10^9 N/m², and density is 1000 kg/m³; see Table 2.1). E_c is composed of the elasticity of the pipe walls and the elasticity of the fluid within. Modulus of elasticity for various materials are

TABLE 2.1
Modulus of Elasticity of Common Pipes and Water

Material	*(10^9) N/m²	*10^6 psi
HDPE	0.8	0.116
Polypropylene	1.5–2	0.22–0.29
Polyethylene terephthalate (PET)	2–2.7	0.29–0.39
Polystyrene	3–3.5	0.44–0.51
Medium-density fiberboard (MDF)	4	0.58
Wood	11	1.60
Glass-reinforced polyester matrix	17.2	2.49
Aromatic peptide nanotubes	19–27	2.76–3.92
High-strength concrete	30	4.35
Carbon fiber-reinforced plastic (50/50 fiber/matrix, biaxial fabric)	30–50	4.35–7.25
Hemp fiber	35	5.08
Magnesium metal (Mg)	45	6.53
Glass	50–90	7.25–13.1
Flax fiber	58	8.41
Aluminum	69	10
Aramid	70.5–112.4	10.2–16.3
Cast iron	110	16
Bronze	96–120	13.9–17.4
Brass	100–125	14.5–18.1
Titanium (Ti)	110.3	16
Copper (Cu)	117	17
Carbon fiber-reinforced plastic (70/30 fiber/matrix, unidirectional)	181	26.3
Wrought iron	190–210	27.6–30.5
Steel (ASTM-A36)	**190**	**28**
Polyvinyl chloride (PVC)	**2.8**	**0.40**
Water	2.2	0.32
Stainless steel	184	27

shown in Table 2.1. Modulus of elasticity is also referred to as Young's Modulus. The combined modulus of elasticity (E_c) may be calculated by the following relationship:

$$\frac{1}{E_c} = \frac{1}{E_b} + \frac{Dk}{E_p.th}$$

(2.5)

In Equation (2.5), E_c is the combined modulus of elasticity, E_b is the modulus of elasticity of water (2.2×10^9 N/m²), E_p is the modulus of elasticity of pipe (Table 2.1), "th" is the pipe wall thickness, D is the inside diameter of the pipe and k is a coefficient calculated from the following rules:

$k = (5/4 - \epsilon)$, for pipes free to move longitudinally,
$k = (1 - \epsilon^2)$, for pipes anchored at both ends against longitudinal movement,
$k = (l - 0.56\,\epsilon)$, for pipes with expansion joints,

where ϵ is the Poisson's ratio of the pipe wall material. It may take on the value $\epsilon = 0.25 - 0.33$ for common pipe materials. *Poisson's ratio is the ratio of the proportional decrease in a lateral measurement to the proportional increase in length in a sample of material that is elastically stretched.* In case of uncertainty, ϵ value of 0.25 can be used for a conservative estimate.

If the longitudinal stress in a pipe is neglected, then $k = 1.0$, and Equation (2.5) can be simplified to

$$\frac{1}{E_c} = \frac{1}{E_b} + \frac{D}{E_p \text{th}} \qquad (2.6)$$

If the time of closure of the valve t is less than $2L/C$, the valve closure is completed before the first pressure wave can return to the valve, and the resulting pressure rise would be the same as if the valve had an instantaneous closure. However, if t is greater than $2L/C$, then the first pressure wave returns to the valve before the valve is completely closed. The returned negative pressure wave compensates the pressure rise resulting from the final closure of the valve.

The maximum pressure created by the water hammer phenomenon may be calculated. The formulas for water hammer pressure are derived as follows:

$$\Delta P = \frac{E_c}{AL}\left[V_0\, A\left(\frac{L}{C}\right)\right] = \frac{E_c V_0}{C} = \rho.C.dV = \rho.C.V \qquad (2.7)$$

which is the pressure head caused by the water hammer. The formula is applicable to rapid valve movement ($t \le 2L/C$) L is defined as the distance from the valve to a relief point such as tank or reservoir.

For closure time $t > 2L/C$, ΔP will not develop fully because the reflected negative wave arriving at the valve will compensate for the pressure rise. For such slow valve closures, the maximum water hammer pressure may be calculated by the Allievi formula.

The maximum water hammer pressure calculated by the Allievi formula is

$$\Delta P = P_0 \left[\frac{N}{2} + \sqrt{\frac{N^2}{4} + N} \; \right] \qquad (2.8)$$

where P_0 is the static state pressure in the pipe, and

$$N = \left(\frac{\rho L V_0}{P_0 t} \right)^2 \qquad (2.9)$$

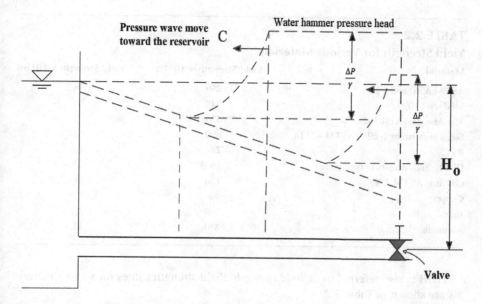

FIGURE 2.2   Water hammer pressure in a pipeline.

The total pressure experienced by the pipe is

$$P = \Delta P + P_0 \qquad (2.10)$$

As the pressure wave travels up the pipeline, energy is being stored in the form of pressure energy in the pipe behind the wave front. Maximum pressure is reached when the wave front arrives at the reservoir:

$$P_{\max} = \gamma H_0 + \Delta P, \qquad (2.10)$$

where H_0 is the total energy head prior to the valve closure, as indicated by the water elevation in the reservoir in Figure 2.2 and γ is the specific weight of water.

2.3.1   CALCULATING PRESSURE TOLERANCE

Pressure tolerance of a pipe can be calculated as follows:

$$P = \frac{HDB}{SDR - 1} \qquad (2.11)$$

where
   HDB = Hydrostatic design basis/tensile strength (35,000 psi for steel, 4,000 psi for PVC)

TABLE 2.2
Yield Strength for Various Materials

Material	Yield Strength × 10⁶ Pa	Yield Strength × 10³ psi
ASTM A36 steel	**250**	**36.2**
Mild steel 1090	248	36
Stainless steel AISI 302	520	75
Steel, high-strength alloy ASTM A5114	690	100
PVC	**28**	**4.0**
HDPE (high-density polyethylene)	26–33	3.8–4.8
Cast iron ASTM A-48	130	19
Copper	70	110
Brass	200	29
Aluminum	240	35

HDB is also referred to as *yield strength*. Yield strength values for various materials are shown in Table 2.2.

P = pressure rating, psi/Pascal

SDR = standard dimension ratio

$$SDR = \frac{OD}{th} \tag{2.12}$$

OD = average outside diameter

th = wall thickness

2.3.2 EXAMPLES

The following examples show how water hammer can be calculated.

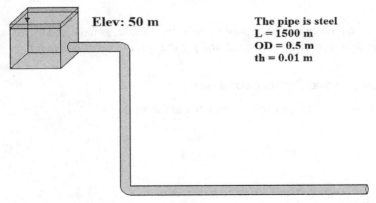

Elev: 50 m

The pipe is steel
L = 1500 m
OD = 0.5 m
th = 0.01 m

Elev: 0.0

Example 2.1

The pipeline shown in Example 2.1 is delivering water from a tank to a community 1500 m away.

The exit pipe connection is rectangular, the elbows are 90 degrees regular flanged and the valve is a 20-inch gate valve. Ignoring the minor losses,

a. Determine the flow rate in m³/s.
b. If the valve is closed within 1.4 sec, calculate the maximum water hammer and the maximum pressure.
c. Determine the pipe tolerance.
d. If the valve is closed in 10 sec, calculate the maximum pressure.
e. If the pipe in part "b" cannot tolerate the pressure, what will happen?

Solution:

a. To calculate the flow and velocity we can use the Hazen–William equation:

$$V = 0.85 C_{HW} R_h^{0.63} S^{0.54}$$

$$V = 0.85(150)\left(\frac{0.5}{4}\right)^{0.63}\left(\frac{50}{1500}\right)^{0.54} = 5.48 \text{ m/s}$$

$$\frac{1}{E_c} = \frac{1}{E_b} + \frac{D}{E_p th}$$

where $E_b = 2.2 \times 10^9$ N/m², and $E_p = 1.9 \times 10^{11}$ N/m², as shown in Table 2.1. The above equation may thus be written as

$$\frac{1}{E_c} = \frac{1}{2.2\times10^9} + \frac{0.5}{(1.9\times10^{11})\times0.01}$$

Hence

$$E_c = 1.39 \times 109 \text{N/m}^2$$

From Equation (2.4), we have the speed of wave propagation along the pipe

$$C = \sqrt{\frac{E_c}{\rho}} = \sqrt{\frac{1.39\times10^9}{1000}} = 1179 \text{m/sec}$$

The time required for the wave to return to the valve is t

$$t = \frac{2L}{C} = \frac{2\times1500}{1179} = 2.54 \text{sec}$$

b. If the valve closes in 1.4 sec, which is less than 2.54 sec, then:
Since 1.4 < 2.54, the water hammer pressure is

$$\Delta P = \rho V_0 C$$

Since the water velocity in the pipe before valve closure is

$$V_0 = 5.48\,\text{m/sec}$$

The maximum water hammer pressure at the valve can be calculated.

$$\Delta P = \rho V_0 C = 1000 \times 5.48 \times 1179 = 6.5 \times 10^6\,\text{N/m}^2$$

The total pressure created in the pipe will become

$$\text{Total } P = P0 + \Delta P = 6.5 \times 10^6 + 1000 \times 9.81 \times 50 = 6.99 \times 10^6\,\text{N/m}^2$$

Therefore the total pressure created in the pipe is equal to 1011.8 psi.

c. Will the pipe tolerate this pressure?
Hydrostatic Design Basis for steel is = 35,000 psi.

$$\text{SDR} = \text{OD}/\text{th} = 0.5/0.01 = 50$$

$$\text{Pt} = \text{HDB}/(\text{SDR} - 1) = 35,000/(50 - 1) = 714\,\text{psi}$$

Pt < 1011.8 psi, the pipe will not tolerate the pressure.
What options do we have?

   1. Slow opening/closing of valve.
   2. Lower the velocity.

d. If the valve is closed in 10 sec, what will be the water hammer?
From Equation (2.9)

$$N = \left(\frac{1000 \times 1500 \times 5.48}{9810 \times 50 \times 10}\right)^2 = 2.81$$

And from Equation (2.8)

$$\Delta P = 0.49 \times 10^6 \left(\frac{2.81}{2} + \sqrt{2.81^2/4 + 2.81}\right) = 1.76 \times 10^6\,\text{N/m}^2 = 254\,\text{psi}$$

$$P_0 = 50 \times 9.81 \times 1000 = 4.9 \times 10^6\,\text{N/m}^2 = 71\,\text{psi}$$

Total P = 71 + 254 = 325 psi which is less than the pipe pressure tolerance of 714 psi

Example 2.2

A cast-iron pipe with diameter of 20 cm and 15-mm wall thickness is carrying water when the outlet is suddenly closed. If the design discharge is 40 liters/sec, calculate the water hammer pressure rise if

(a) The pipe wall is rigid;
(b) The longitudinal stress is neglected;
(c) The pipeline has expansion joints throughout its length.

Solution:

$$A = \frac{\pi}{4} \times 0.22 = 0.0314 \, m^2$$

Hence,

$$V = \frac{Q}{A} = \frac{0.04}{0.0314} = 1.274 \, m/sec$$

(a) For rigid pipe wall, $Dk/E_p.th = 0$, Equation (2.5) gives the following relation

$$\frac{1}{E_c} = \frac{1}{E_b} \text{ or } E_c = E_b = 2.2 \times 109 \, N/m^2$$

From Equation (2.4), we can calculate speed of pressure wave,

$$C = \sqrt{\frac{E_c}{\rho}} = \sqrt{\frac{2.2 \times 10^9}{1000}} = 1483 \, m/sec$$

and the water hammer pressure is

$$\Delta P = \rho.V.C = 1000(1.274)*(1483) = 1.89 \times 10^6 N/m^2$$

$$\rightarrow P = 1.89 \times 10^6 N/m^2 = 273 \, psi$$

(b) For pipes with no longitudinal stress, $k = 1$, we may use Equation (2.3b)

$$E_c = \frac{1}{\left(\dfrac{1}{E_b} + \dfrac{D}{E_p th}\right)} = \frac{1}{\left(\dfrac{1}{2.2 \times 10^9} + \dfrac{0.2}{(1.6 \times 10^{11})(0.015)}\right)} = 1.86 \times 10^9$$

and

$$C = \sqrt{\frac{E_c}{\rho}} = 1364 \,\text{m/sec}$$

Hence, the rise of water hammer pressure can be calculated:

$$\Delta P = \rho.V.C = (1000)(1.274)(1364) = 1.74 \times 10^6 \,\text{N/m}^2 = 251 \,\text{psi}$$

(c) For pipes with expansion joints, $k = (1 - 0.5 \times 0.25) = 0.875$

$$E_c = \cfrac{1}{\left(\cfrac{1}{E_b} + \cfrac{0.875D}{E_p th}\right)} = \cfrac{1}{\left(\cfrac{1}{2.2 \times 10^9} + \cfrac{0.875(0.2)}{(1.6 \times 10^{11})(0.015)}\right)} = 1.90 \times 10^9 \,\text{N/m}^2$$

and

$$C = \sqrt{\frac{E_c}{\rho}} = 1378 \,\text{m/sec}$$

Equation (2.7) is used to calculate the rise in water hammer pressure.

$$\Delta P = \rho.V.C = 1000(1.274)(1378) = 1.76 \times 10^6 \,\text{N/m}^2$$

$$\rightarrow P = 254 \,\text{psi}$$

Example 2.3: PVC pipe pressure tolerance

Establish the pressure rating of 8-inch schedule 40 PVC pipe (polyvinyl chloride).
From Table 2.5, based on OD and pipe thickness,

OD = 8.625 in. (8-inch nominal iron pipe size (IPS))
th = 0.322 in.
HDB = 4000 psi

Solution:

Calculation of SDR

$$SDR = \frac{OD}{th} = \frac{8.625}{0.322} = 26.8$$

Calculation of pressure rating (Pt)

$$Pt = \frac{HDB}{SDR - 1} = \frac{4000}{26.8 - 1} = 155 \,\text{psi}$$

The calculations show that the 8-inch schedule 40 PVC pipe has a rating of 155 psi. Note that Table 2.5 shows a rating of 160 psi.

The pressure ratings in Table 2.5 are for an operating temperature of 73.4 F (23°C). When a PVC pressure pipe operates at temperatures other than 73.4 F (23°C), pressure capacity should be corrected based on the thermal correction factors shown in Table 2.3.

TABLE 2.3
Temperature Pressure Correction Factors

Temp. °F	Correction Factor
80	0.88
90	0.75
100	0.62
110	0.50
120	0.40
130	0.30
140	0.22

TABLE 2.4
Schedule 40 Standard Steel Pipe Dimensions

Size in	Outside Diameter in	Inside diameter in	Wall thickness in	Weight per ft. lb
1/4	0.540	0.364	0.0880	0.424
3/8	0.675	0.493	0.0910	0.567
1/2	0.840	0.622	0.1090	0.850
3/4	1.05	0.824	0.1130	1.13
1	1.315	1.049	0.1330	1.678
1 1/2	1.660	1.380	0.1400	2.272
1 1/2	1.900	1.610	0.1450	2.717
2	2.375	2.067	0.1540	3.652
2 1/2	2.875	2.469	0.2030	5.793
3	3.500	3.068	0.2160	7.575
3 1/2	4.000	3.548	0.2260	9.109
4	4.500	4.026	0.2370	10.790
5	5.563	5.047	0.2580	14.617
6	6.625	6.065	0.2800	18.974
8	8.625	7.981	0.3220	28.554
10	10.750	10.020	0.3650	40.483
11	11.750	11.000	0.3750	45.557
12	12.750	12.000	0.3750	49.562

TABLE 2.5
Schedule 40 PVC Pipe Dimensions

Nom. Pipe Size (in)	O.D. in	Average I.D. in	Min. Wall thickness in	Nominal Wt./Ft. lb	Maximum W.P. psi*
1/8	0.405	0.249	0.068	0.051	810
1/4	0.540	0.344	0.088	0.086	780
3/8	0.675	0.473	0.091	0.115	620
1/2	0.840	0.602	0.109	0.170	600
3/4	1.050	0.804	0.113	0.226	480
1	1.315	1.029	0.133	0.333	450
1 1/4	1.660	1.360	0.140	0.450	370
1 1/2	1.900	1.590	0.145	0.537	330
2	2.375	2.047	0.154	0.720	280
2 1/2	2.875	2.445	0.203	1.136	300
3	3.500	3.042	0.216	1.488	260
3 1/2	4.000	3.521	0.226	1.789	240
4	4.500	3.998	0.237	2.118	220
5	5.563	5.016	0.258	2.874	190
6	6.625	6.031	0.280	3.733	180
8	8.625	7.942	0.322	5.619	160
10	10.750	9.976	0.365	7.966	140
12	12.750	11.889	0.406	10.534	130
14	14.000	13.073	0.437	12.462	130
16	16.000	14.940	0.500	16.286	130
18	18.000	16.809	0.562	20.587	130
20	20.000	18.743	0.593	24.183	120
24	24.000	22.544	0.687	33.652	120

For the exampe above, the pressure tolerance of 8-inch SCH 40 PVC was calculated as 155 psi. If the temperature of the liquid is changed to 100°F, then the pressure tolerance will be

$$P = 155*(0.62) = 96.1\,\text{psi}$$

2.3.3 PREVENTIVE AND MITIGATING MEASURES FOR PIPE FAILURE DUE TO WATER HAMMER

1. Positive Water Hammer, $\Delta P = \rho. C. dV = \rho. C. V$
   a. Reduce V
   b. Reduce dv, by slow opening/slow closing
   c. Air vent, prevent air entrapment
   d. Stand pipe

2. Negative Water Hammer, $\Delta P = -\rho . C . dV = -\rho . C . V$
   a. Reduce V
   b. Reduce dv by slow opening/slow closing
   c. Air Vent, Prevent negative pressure
   d. Stand pipe
3. Mitigating Measures
   a. Increased thickness (expensive)
   b. Increased tensile strength, steel versus PVC (more expensive)
   c. Pressure relief valve, PRV.

2.4 CALCULATING COLLAPSE STRENGTH

2.4.1 COLLAPSE STRENGTH FOR PVC

A pipe can collapse due to negative internal pressure and positive external load if it exceeds the collapse tolerance (P_c) of the pipe. Pipe collapse tolerance can be calculated using a modified Timoshenko's equation.

The following is the ASTM F 480 formula for determining the collapse strength of a PVC pipe manufactured in accordance with the said standards.

$$P_c = \left(\frac{2E}{1-u^2} \right) \left\{ \frac{1}{\left[\dfrac{D_0}{th} \right]\left[\dfrac{D_0}{th} - 1 \right]^2} \right\} \qquad (2.13)$$

where:

P_c = Collapse pressure tolerance of PVC pipe (psi)
E = Young's Modulus for PVC (4×10^5 psi)
u = Poisson's ratio (0.25–0.33)
D_0 = Outside diameter of pipe (inches)
th = Wall thickness (inches)

Another formula used to determine the collapse strength of PVC is shown below. This formula is more conservative than the ASTM formula. However, since most PVC materials are made to minimum allowable wall thickness rather than nominal sizes, it may provide a more accurate estimate of actual collapse strength.

$$P_c = \frac{(2E)0.75}{(1-u^2)\left[\dfrac{D_0}{th} - 1 \right]^3} \qquad (2.14)$$

Equation (2.14) is a conservative version of Timoshenko's Equation (2.16). Physical properties of PVC vary with temperature. The values obtained with these formulas are consistent with a temperature of 70°F. As the temperature rises, PVC working strength decreases by approximately 0.5 psi per degree Fahrenheit above 70°F. Obviously, much care must be taken during cementing operations or in other high-temperature environments.

2.4.2 COLLAPSE STRENGTH OF STEEL CASING

Timoshenko's formula, shown below, is commonly accepted as the most accurate method available for estimating the collapse strength of steel pipes with diameter/thickness ratios common to those used in the construction of water wells.

$$
P_e^2 - P_e \left\{ \frac{25}{\left[\frac{D_0}{\text{th}} - 1 \right]} + P_{cr} \left(1 + 3e \left[\frac{D_0}{\text{th}} - 1 \right] \right) \right\} + \left\{ \frac{2SP_{cr}}{\left[\frac{D_0}{\text{th}} - 1 \right]} \right\} = 0 \qquad (2.15)
$$

where
   P_{cr} = Theoretical collapse strength of a perfectly round tube (psi)

$$
P_{cr} = \left(\frac{2E}{1 - u^2} \right) \left\{ \frac{1}{\left[\frac{D_0}{\text{th}} - 1 \right]} \right\}^3 \qquad (2.16)
$$

where

   P_e = Collapse pressure with ellipticity (psi)
   E = Young's Modulus (modulus of elasticity) for steel (3×10^7 psi)
   u = Poisson's ratio (0.25–0.33)
   D_0 = Outside diameter of casing (inches)
   e = Ellipticity, usually taken as 0.01
   S = Tensile strength (35,000 psi for steel, 4000 psi for PVC)

* Poisson's ratio is the ratio of the proportional decrease in a lateral measurement to the proportional increase in length in a sample of material that is elastically stretched. In the absence of a measured Poisson's ratio, use a conservative value of 0.25.

2.4.3 BOUSSINESQ'S FORMULA FOR POINT LOADS

Figure 2.3 shows a load Q acting at a point P on the surface of a semi-infinite solid. A semi-infinite solid is one bounded on one side by a horizontal surface, here the

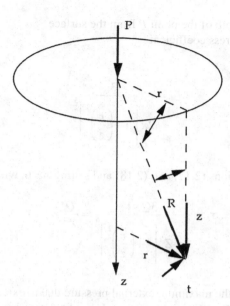

FIGURE 2.3 Symbols used in the equations for a point load applied at the surface.

surface of the earth, and infinite in all the other directions. The problem of determining stresses at any point P at a depth Z as a result of a surface point load was solved by Boussinesq (1885) on the following assumptions:

1. The soil mass is elastic, isotropic, homogeneous, and semi-infinite.
2. The soil is weightless.
3. The load is a point load acting on the surface.

The soil is said to be isotropic if there are identical elastic properties throughout the mass and in every direction through any point of it. The soil is said to be homogeneous if there are identical elastic properties at every point of the mass in identical directions. The expression obtained by Boussinesq for computing vertical stress (σ_z) at depth Z (Figure 2.3) due to a point load Q is

$$\sigma_z = \frac{3Q \times 1}{2\pi Z^2 \left[1 + \left(\dfrac{r}{z} \right)^2 \right]^{\frac{5}{2}}} = \frac{Q}{Z^2} \times I_B \tag{2.17}$$

where

r = the horizontal distance between an arbitrary point below the surface and the vertical axis through the point P

z = the vertical depth of the point P from the surface
I_B = Boussinesq stress coefficient

$$I_B = \frac{3}{2\pi \left[1 + \left(\dfrac{r}{z} \right)^2 \right]^{\frac{5}{2}}} \qquad (2.18)$$

Combining Equations (2.17) and (2.18) and setting $r = 0$, will result in

$$\sigma_z = \frac{3Q \times 1}{2\pi\,Z^2 \left[1 + \left(\dfrac{r}{z} \right)^2 \right]^{\frac{5}{2}}} = \frac{Q}{Z^2} \times \frac{3}{2\pi} \qquad (2.19)$$

Where σ_z represents the maximum external pressure due to external load Q.

The values of the Boussinesq coefficient I_B can be determined for different values of r/z. The variation of I_B with r/z in a graphical form is given in Figure 2.4. The maximum stress due to point load Q at depth Z can occur when r in Equation (2.19) is equal to zero.

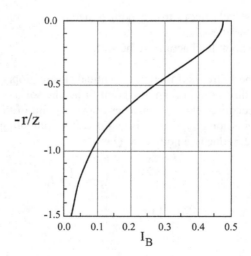

FIGURE 2.4 The influence factor as a function of the ratio (r/z) from a vertical point load at the surface.

Example 2.4

a. Calculate collapse strength (tolerance) for 8-inch schedule 40 PVC pipe, with 0.322-inch wall thickness.

Solution:

The collapse pressure tolerance would be

$$P_c = \frac{(2E)0.75}{\left(1-0.33^2\right)\left[\dfrac{8.625}{0.322}-1\right]^3} = 39.27 \text{ psi}$$

b. Determine the external pressure on the 8-inch pipe if buried 0.3 m below the surface, and a point load of 8 tons is applied on the surface.

Solution:

The maximum external pressure occurs when $r = 0$,

$$r = 0.0$$

$$I_B = 3/(2\pi) = 0.477$$

$$\delta = \frac{Q}{Z^2} \times I_B = \frac{8000 \times 9.81}{0.3^2} \times 0.47 = 409 \times 10^3 \text{N}/\text{m}^2$$

$= 59.2$ psi > 39.27, therefore the pipe will collapse due to external pressure.

c. Would the pipe be safe if it is buried 0.6 m below the surface?

Solution:

$$\delta = \frac{Q}{Z^2} \times I_B = \frac{8000 \times 9.81}{0.6^2} \times 0.47 = 102 \times 10^3 \text{N}/\text{m}^2$$

$= 14.8$ psi < 39.27, the pipe will be safe.

It is common practice to bury the pipe at a certain depth to prevent collapse due to external load and freeze protection. Table 2.4 shows the depths for various PVC pipes recommended for collapse prevention by the Natural Resources Conservation Service (USDA/NRCS).

2.4.4   PREVENTIVE MEASURES TO AVOID PIPE COLLAPSE

a. Air vent
b. Slow opening – Slow closing
c. Thicker pipe (expensive)
d. Higher Modulus of Elasticity (expensive)
e. Bury the pipe deeper (costly)

2.5   FAILURE DUE TO FREEZING

A pipe may fail due to freezing. The volume of water expands when it freezes. The relationship between change in pressure due to change in volume is described below.

$$E_b = \frac{dP}{\dfrac{dV}{V}} \qquad\qquad (2.20)$$

where
   dP = change in pressure
   dV = change in volume due to freezing
   V = volume

Example 2.5

When water freezes, the volume will increase by about 8%, $\dfrac{dV}{V} = 0.08$. Calculate the change in pressure in the pipe.

Solution:

$$E_b = 2.2 \times 10^9 \, \text{N/m}^2$$

Using Equation (2.19),

$$dP = (2.2) \times 10^9 \times (0.08) = 17.6 \times 10^7 \, \text{N/m}^2 = 25,475 \, \text{psi}$$

Freezing can result in significant increase in pressure that can burst the pipe. Preventive measures such as draining the pipe and burying it below the freeze line are critical in preventing costly repair.

2.5.1   PREVENTIVE MEASURES FOR FREEZING

   a. Bury the pipe deeper
   b. Drain the pipe
   c. Insulate the pipe

2.6   PREVENTIVE AND MITIGATING MEASURES FOR PIPE SAFETY

Pipes will fail due to (a) excessive internal pressure, (b) excessive collapse pressure.

2.6.1   EXCESSIVE INTERNAL PRESSURE

The following preventive steps can be taken:

   1. Slow opening – Slow closing
   2. Air valve

3. Drain valve
4. Stand pipe
5. Low velocity

The following mitigating measures can be taken:

1. Thicker pipe
2. Stand pipe
3. Pressure relief valve
4. Air valve
5. Surge tank
6. Harder material

Figure 2.5 shows an air valve and a pressure relief valve (PRV) installed downstream of a pump and upstream of a buried pipeline. The purpose of the air valve is to prevent negative pressure when the system is turned off suddenly. The purpose of a PRV is to prevent the pipe breaking due to excessive positive water hammer which can occur if the system downstream of the valve is closed or closed suddenly.

FIGURE 2.5   Air valve and PRV installed downstream of a pump.

2.6.2 EXCESSIVE COLLAPSE PRESSURE

The following preventive measures can be taken against negative pressure:

1. Air vent
2. Slow-opening and slow-closing valve

The following preventive measures due to point load can be taken:

1. Bury the pipe deeper
2. Use stronger pipe

A common method to avoid sending positive or negative water hammer into a pipe network is through hydraulic valves (Figures 2.7 and 2.8. Hydraulic valves are two-chamber valves where hydraulic pressure through a solenoid valve is used to open or close the main valve slowly, thus preventing strong water hammer.

Figure 2.5 shows air valve and pressure relief valve (PRV) downstream of a pump and a valve to prevent and mitigate positive and negative water hammer. The PRV was rated at 50 psi which was equal to the pressure tolerance of the pipe.

Figure 2.6 demonstrates the application of a parallel discharge system downstream of a pump station to prevent water hammer. The system is installed at the outlet of a pump station that is pumping 1500 gpm into a pressurized municipal pipe

FIGURE 2.6 Parallel discharge system for slow opening/closing downstream of a pump.

network system. Prior to starting the pump, the valve to the network is closed and the valve to a free discharge outlet is fully open. After the pump is turned on, the open discharge valve is slowly closed using a double-chamber hydraulic valve and the valve leading to the network slowly opens due to increasing back pressure. After a short time, the open discharge valve is fully closed and the network valve is fully open, preventing water hammer. The pressure transition takes about 20–30 seconds. The same process is applied in reverse before the pump is turned off to avoid occurrence of severe negative water hammer.

The mechanism of operation of hydraulic valves is demonstrated in Figures 2.7 and 2.8.

Figure 2.9 shows a manual valve downstream of a pump where a slow-opening, slow-closing valve is used to avoid water hammer.

The water supply system should allow a distance of at least 30 ft between the pump and where the line goes underground or connects to distribution lines, for instrumentation Figure 2.10.

Flow meters (water meters) use multi-blade propellers whose speed of rotation is determined by the speed of the water flow. Thus, the propeller measures water velocity in the pipeline, and the indicator converts this measurement to flow rate based on the inside diameter of the pipeline.

Proper installation of propeller flow meters requires the following:

1. The proper selection of the meter based on pipe size, range of flow and head loss.
2. Installation of the flow meter so the pipe is always flowing full.
3. The flow meter should be installed at a specific distance from the closest pipe angle (normally 4 times the pipe diameter).

Discharge pipe controlled by solenoid valve releases the pressure from the upper chamber and
Opens the valve

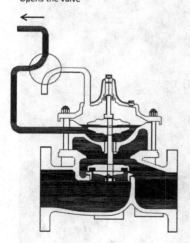

FIGURE 2.7 (open valve). The upper chamber is empty and water flows through the valve by pushing up the piston.

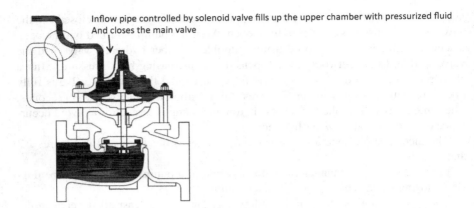

Inflow pipe controlled by solenoid valve fills up the upper chamber with pressurized fluid
And closes the main valve

FIGURE 2.8 (closed valve). The upper chamber is pressurized and the hydraulic pressure pushes the piston down and closes the flow path.

FIGURE 2.9 Slow-opening and slow-closing manual valve installed downstream of a pump.
Source: New Mexico State University Hydraulic Laboratory.

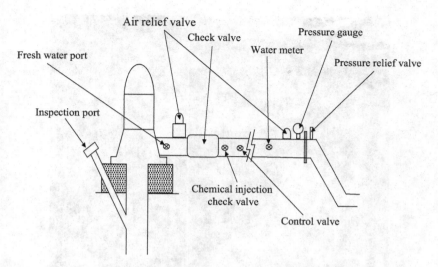

FIGURE 2.10 Instrumentation in a hydraulic system consisting of a pump and an underground pipe delivering water into a center pivot system.

Pipes are normally buried to prevent collapse due to external load and freezing. Burying the pipe deeper will reduce the impact of external load and prevent pipe collapse, as shown in Example 2.4.

Table 2.6 describes the minimum cover for PVC pipes given by NRCS (Natural Resources Conservation Service) in order to prevent collapse due to external load.

Pipes' tolerance to negative pressure is normally lower than burst pressure. Negative pressure will result in pipe collapse, especially in thin or plastic pipes. Figure 2.11 shows a collapsed buried PVC pipe due to an unsuitable air valve at the pump. Note the pipe is warped at the junction.

TABLE 2.6
NRCS Standard and Specs for Buried PVC Pipes

The minimum depth of cover shall be:

No	Diameter (inches)	Depth of Cover (inches)
1	1/2 through $2\frac{1}{2}$	18
2	3 through 5	24
3	6 through 18	30
4	Greater than 18	36
5	All sizes in soils subject to cracking	36

FIGURE 2.11   Pipe collapsed due to unsuitable air vent at the pump.

REFERENCE

Timoshenko, S. and Gere, J. 1961. *Theory of Elastic Stability*, McGraw Hill, New York.

3 Design of Pumping Systems

Kinetic Pumps; Vertical turbine, submersible, & Centrifugal

3.1 SUMMARY EQUATIONS

Energy equation:

$$E_1 = E_2 + h_L - h_p,$$ (3.1)

Or

$$E_1 = E_2 + h_L - \text{TDH},$$ (3.2)

Water horsepower, $\text{WHP} = \dfrac{Q \times \text{TDH}}{3960}$, Q in gpm, TDH in ft,

Brake horsepower, $\text{BHP} = \dfrac{\text{WHP}}{E_P}$ or $\text{BHP} = \dfrac{Q \times \text{TDH}}{3960 \times E_P}$,

Motor horsepower, $\text{MHP} = \dfrac{\text{BHP}}{E_s}$, E_s is shaft efficiency

Input energy $= \dfrac{\text{MHP}}{E_m}$, E_m = motor efficiency

Maximum suction lift $= H_S = H_0 - H_V - \text{NPSHR} - H_f$

DOI: 10.1201/9781003287537-3

57

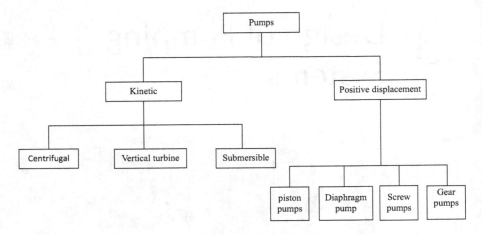

FIGURE 3.1   Classification of pumps.

3.2   CLASSIFICATION OF PUMPS

Pumps may be classified based on the application they serve, the materials from which they are constructed, the liquids they handle, and even their orientation in space. All such classifications, however, are limited in scope and tend to substantially overlap each other. Another system of classification could be based on the principles by which energy is added to the system and the means by which this principle is implemented. The type of energy added to the system could also be variable. The classification shown here is related to the pump itself and is unrelated to any consideration external to the pump or even to the materials from which it may be constructed. Figure 3.1 shows the general classification of pumps.

3.3   PUMP DESIGN

Design of the pump starts with the energy conservation equation:

$$E_1 = E_2 + h_L - h_p \tag{3.1}$$

or

$$E_1 = E_2 + h_L - \text{TDH}, \tag{3.2}$$

Some sources describe the term TDH (total dynamic head) as h_p

where h_p = TDH = pump head
and

$$h_L = \text{head loss} = h_f + h_m$$

Pump selection should be such that the pump can produce the head (h_p) and flow required by the system.

3.3.1 PUMP DESIGN CRITERIA

1. The pump should produce the desired head and flow (meet the objective).
2. The pump should operate at the right-hand side of peak efficiency.
3. The pump should be economical (lowest cost).

3.3.2 PUMP EQUATIONS

Figure 3.2, together with Equations (3.3–3.6), shows various components of a pumping system.

In a conventional system where Q is in gpm, and TDH is in ft, the power in units of water horsepower (WHP) is calculated as:

$$\text{WHP} = \frac{Q \times \text{TDH}}{3960} \tag{3.3a}$$

In the metric system, where Q is in m³/s, and TDH is in meters, the power in units of N-M/s or watts is calculated as

$$\text{Power} = \gamma \cdot Q \cdot \text{TDH} \tag{3.3b}$$

and input power to the pump, IW, is calculated as

$$\text{IW} = \gamma \cdot Q \cdot \text{TDH}/E_p$$

where IW is input power to the pump, γ is specific weight of water (9810 N/m³) and E_p is pump efficiency in percentage/decimal.

In the English system where Q is in cfs, and head is in ft., the power in Equation (3.3b) would be in lb-ft/sec. lb-ft/sec can be converted to horsepower (HP) by dividing the value by 549.23 (550).

In conventional units, where Q is in gpm, and TDH is in ft, the input power to the pump is called brake horsepower (BHP) and is calculated as:

$$\text{BHP} = \frac{\text{WHP}}{E_p} \quad \text{or} \quad \text{BHP} = \frac{Q \times \text{TDH}}{3960 \times E_p} \tag{3.4}$$

where E_p = pump efficiency
and motor horsepower (output power from the motor or engine) is

$$\text{MHP} = \frac{\text{BHP}}{E_s} \tag{3.5}$$

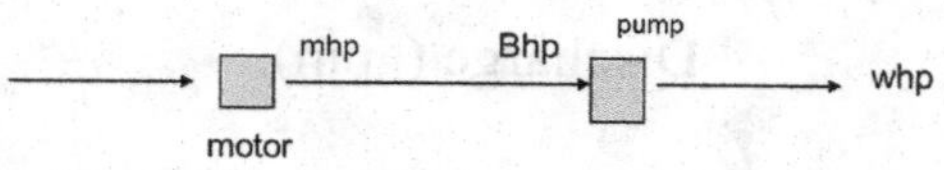

FIGURE 3.2 Pumping system components.

where E_s = shaft efficiency. In close-coupled units such as centrifugal and submersible pumps, E_s is assumed equal to one since the energy loss in the shaft is negligible. Input energy to the motor or engine is calculated as

$$\text{Input energy} = \frac{\text{MHP}}{E_m}, \tag{3.6}$$

where E_m = motor efficiency

3.3.3 PUMP CHARACTERISTIC CURVES

A typical pump has four curves. These are

a. Head discharge curve
b. Efficiency curve
c. BHP curve
d. NPSH (net positive suction head) curve

Figure 3.3 shows various characteristic curves for a typical pump. In Figure 3.3, the system curve is not a typical pump characteristic curve.

The following examples describe the design of a pump and motor system for various applications.

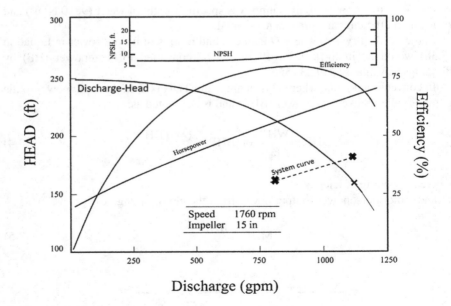

FIGURE 3.3 Characteristic curves in a typical pump. The system curve is not part of the pump information.

Example 3.1a: Design of a centrifugal pump

Water is being pumped from a reservoir to a tank. The desired pumping rate is 1100 gpm. Water level elevation in the reservoir is equal to 25 ft and water level elevation in the tank is at 142 ft. The pipe is schedule 40 PVC. Ignoring the minor losses, determine:

1. Pipe size
2. Design point
3. Motor size
4. Input energy
5. Energy cost

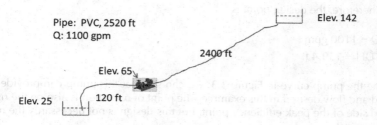

Pipe: PVC, 2520 ft
Q: 1100 gpm

2400 ft

Elev. 142

Elev. 65

Elev. 25   120 ft

Pipe Length:
Reservoir 1 to pump: 120 ft
Pump to reservoir 2: 2400 ft

Using Equation (3.1)

$$E_1 = E_2 + h_L - h_p$$

$$0 + 0 + 25 = 0 + 0 + 142 + h_L - h_p$$

thus

$$h_p = 117 + h_L$$

Ignoring minor losses, $h_L = h_f$ can be calculated from H–W equation
However, prior to calculation of h_f, we need to decide on the design of the pipe.
Using a guideline of 7–10 ft/sec,

$$V.A = Q$$

$$7 \times \left(\frac{\pi.D^2}{4} \right) = \frac{1100}{449}$$

and from here $D = 0.668$ ft $= 8$ inches.
Using 8 inch in H–W equation, friction loss (h_f) can be calculated as

$$h_f = 10.5(2520) \times \left(\frac{1100}{150}\right)^{1.852} \times (8)^{-4.87}$$

$$h_f = 42.37\,\text{ft}$$

and

$$h_p = 117 + 42.37 = 159.37\,\text{ft}$$

Therefore, the design point is

$Q = 1100$ gpm
TDH $= 159.4$ ft

Using the pump curve in Figure 3.3, we can see that this pump can provide the head and flow desired in this example. The point of operation is also on the right-hand side of the peak efficiency point, but this design is not necessarily the most economical.

The design also shows that this operation point has an efficiency of about 73% and NPSH of 20 ft.

In the following calculations, we will see how these values are used.

Referring to Figure 3.3, the BHP (brake horsepower) or the input energy required at the pump is calculated, using Equation (3.4), as:

$$\text{BHP} = \frac{1100 \times 159.34}{3960 \times 0.73} = 60.6\text{HP}$$

and the MHP (motor size) can be calculated as:

$$\text{MHP} = \frac{60.6}{E_s}, \text{ using } E_s \text{ of } 100\%, \text{ or } 1, \text{ MHP} = 60.6\text{HP}$$

In this calculation, the shaft efficiency (E_s) is set equal to 1, (100%) due to the close-coupled nature of the centrifugal pump. The shaft energy loss in close-coupled centrifugal pumps and submersible pumps is generally set equal to zero.

This is the motor size needed for this system. Referring to Table 3.4, the closest motor that can be selected is 75 HP, with motor efficiency of 91%.

$$\text{Since one HP} = 0.746\,\text{kW}$$

the input energy required (if an electric motor is used) is:

$$\text{Input energy} = \frac{60.6}{0.91} = 66.59\text{HP} = 0.746(66.59) = 49.7 \text{ kW}$$

If the pump operates for 24 hours and electricity cost is $0.12/kW h, then the operation cost is:

Cost = (input power)*(T)*($/kW h) = 49.7 (24)*0.12) = $143.14 for 24 hours (from Equation (3.8a), page xx).

An alternative to the pump shown in Figure 3.3 would be a pump shown in Figure 3.4. In this pump curve, the design point of 1100 gpm and h_p of 159.4 ft falls slightly over the pump curve model 3902 and impeller size of 11 inches. The pump efficiency in this case would be 76% (0.76). This will result in MHP of 58.24 HP, and input energy of 64 HP or 47.74 kW. This alternative design can produce the desired Q and h_p with lower energy cost. The NPSH value for this design is 13.75 ft.

Another alternative design would be to select a submersible pump shown in Figure 4.9. In this pump curve, the design point of 1100 gpm, and 159.4 ft of head falls on the pump curve "800S750-3A" with a 75 HP motor. The efficiency of this pump would be 68% (0.68). This option will result in a MHP of 65 HP and input energy of 72.35 HP (54 kW). The NPSH for this design can be taken from Figure 4.12 as 53 cft.

Example 3.1.b

If the design point does not fall on the pump's Q–H curve, then the designer will have three options:

1. Select a different pump.
2. Change the system to fit the pump curve.
3. Find a compromise point using the system curve.

The following example shows how this can be accomplished.
What if the elevation of the reservoir in Example 3.1a was 162 ft?

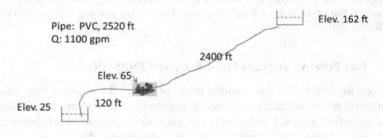

Pipe: PVC, 2520 ft
Q: 1100 gpm

Elev. 162 ft

2400 ft

Elev. 65

120 ft

Elev. 25

Pipe Length:
Reservoir 1 to pump: 120 ft
Pump to reservoir 2: 2400 ft

Then TDH1 = 162 − 25 + 42.37 = 179.37 ft

Note the new design point of (Q = 1100 gpm and TDH = 179.4 ft) falls above the pump curve in Figure 3.3.

We use the option of finding a compromise point by drawing a system curve.

To draw a system curve, we choose a lower flow rate like 800 gpm and re-calculate TDH.

$$TDH2 = 162 - 25 + 23.5 = 160.5 \text{ ft}$$

Then we plot the system curve to find the new operation point (compromise point).

The new operation point will therefore be (Figure 3.3)

Q = 1050 gpm

Head = 170 ft

A second option is to select a vertical turbine pump curve (see Figure 4.3, and combine impellers in series as described later in this chapter. This option will give us:

$6A$ = 6 * 22 = 132

$1B$ = 18

$2C$ = 2 * 15 = 30

Total head = 180 ft

In this case, we used 6 of impeller A, 1 of impeller B and 2 of impeller C installed in series.

In this case, we don't need to draw a system curve, because the design point falls on the combined pump curve.

A third option is to select a submersible pump curve (Figure 4.9), in which the design point of 1100 gpm and head of 179.4 ft falls slightly above the pump curve for which we need to draw a system curve. This option will result in slightly lower pumping rate.

3.3.4 Net Positive Suction Head Required (NPSHR)

Under certain conditions, the absolute pressure within the impeller may fall below the net positive suction head required by the pump. This condition will result in a process called *cavitation*. Cavitation is the process of vapor, air, and bubbles expanding in the suction line due to low pressure and collapsing as the pressure at the pump inlet suddenly increases. The implosion of the air and vapor bubbles results in damage to the impeller assembly, reduced pump efficiency, and reduced flow. A pump that is cavitating will sound as if it is grinding sand.

Cavitation may be avoided by ensuring that the pressure at the pump inlet, called the *net positive suction head available* or NPSHA, meets or exceeds the inlet pressure required by the pump. This required inlet pressure is called the *net positive suction head required* or NPSHR, and is specified by the pump manufacturer.

To avoid cavitation, a pump can be designed such that the maximum suction lift (vertical distance between water level and pump inlet) does not exceed a certain value as calculated by Equation 3.7.

$$H_S = H_0 - H_V - NPSHR - H_f, \tag{3.7}$$

where

H_0 = Atmospheric pressure (feet of water)
H_V = Saturation vapor pressure (feet of water)
H_S = Maximum height of impeller eye above water surface (Ft)
H_f = Friction losses in suction line (ft)

The H_0 depends on elevation above sea level and H_V depends on the water temperature. The values of $(H_0 - H_V)$ are shown in Table 3.1. In Example 3.1a, with the pump located in Las Cruces, NM, at an elevation of 4000 ft above sea level, the NPSH required by the pump at the point of operation was 20 ft, and assuming maximum water temperature of 80°F, then from Table 3.1:

$$H_0 - H_V = 28.3$$

and h_f for the 120 ft of suction line is 2.02 ft (H–W equation), using Equation 3.7,

$$H_S = 28.3 - 20 - 2.02 = 6.28\,ft$$

TABLE 3.1
$(H_0 - H_V)$ for a Range of Temperatures and Elevations

Water Temp. Degrees F	Elevation Above Sea Level (ft)						
	0	1000	2000	3000	4000	5000	6000
30	33.7	32.5	31.3	30.3	29.2	28.1	27
50	33.6	32.3	31.3	30.2	29.1	28	26.9
60	33.3	32.2	31.1	30	28.9	27.8	26.7
70	33.2	32	30.9	29.8	28.7	27.6	26.5
80	32.8	31.6	30.5	29.3	28.3	27.2	26.1
90	32.3	31.2	30.1	29	27.9	26.8	25.7
100	31.8	30.6	29.5	28.3	27.3	26.2	25.1
110	31.1	29.9	28.8	27.7	26.6	25.5	23.3
120	30.1	28.9	27.8	26.7	25.6	23.5	23.3
130	28.9	27.7	26.6	25.5	23.5	23.3	22.2
130	27.3	26.1	25	23.9	22.8	21.7	20.6
150	25.3	23.2	23.1	22	20.9	19.8	18.7

Looking at the diagram in Example 3.1a, the pump is at an elevation of 65 − 25 = 40 ft above the water which far exceeds the maximum allowable suction lift of 6.28 ft calculated.

In this case, the pump needs to be lowered to an elevation below the (25 + 6.28) = 31.28 ft, in order to avoid cavitation which will result in not only damage to the pump impeller but also a reduction in pump efficiency and flow rate.

Alternative Pumps

In this example, if the alternative centrifugal pump of Figure 3.4 is selected, the H_S would be calculated as follows:

$$H_S = 28.3 - 13.75 - 2.02 = 12.53\,\text{ft.}$$

And if the submersible pump of Figure 4.9 is used, the H_S would be calculated as:

$$H_S = 28.3 - 53 - 0.0 = -24.70\,\text{ft.}$$

This means that the pump inlet needs to be submerged 24.7 ft below the water surface.

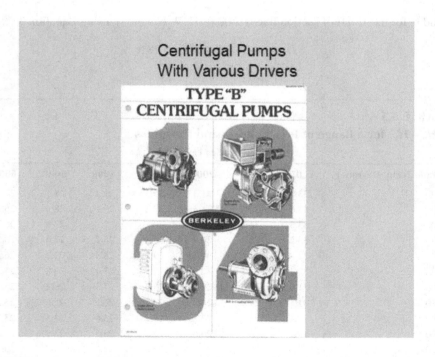

FIGURE 3.4 Typical centrifugal pump curves for various impeller sizes for the same model.

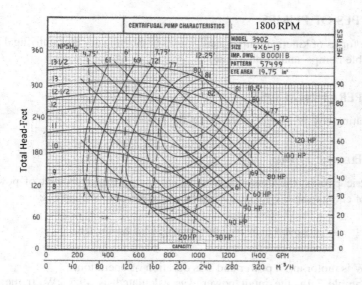

FIGURE 3.4   (Continued)

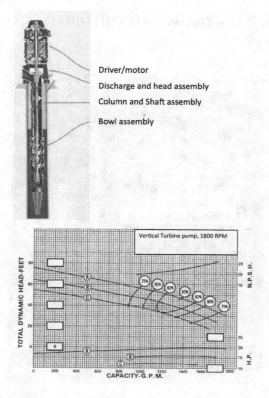

Driver/motor
Discharge and head assembly
Column and Shaft assembly
Bowl assembly

FIGURE 3.5   Submersible pump and typical pump curve.

3.4   TYPES OF PUMPS

The three forms of kinetic pump typically used in the industry are centrifugal pumps, submersible pumps, and turbine pumps.

3.5   OPERATING COST

The operating cost of a pump depends on the power source.

3.5.1   ELECTRIC POWER

For electric pumps, the operating cost is calculated based on cost of power and amount of energy used. The equation for calculating cost of power is:

$$\text{Cost} = \left(\text{kW of input energy}\right)\left(\text{T}\right)\left(\$/\text{kW h}\right), \tag{3.8a}$$

where kW is motor input power and T is duration of operation.

In Example 3.1a, the input power was calculated as 49.7 kW. If the cost of electricity is $0.15/kW h, then the cost of operation for 24 hours is:

$$\text{Cost} / 24\,\text{hrs of operation} = \left(49.7\right)\left(24\right)\left(0.12\right) = \$143.14$$

FIGURE 3.6   Vertical turbine pump and typical impeller curves.

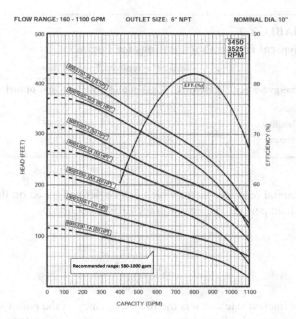

FLOW RANGE: 160 - 1100 GPM   OUTLET SIZE: 6" NPT   NOMINAL DIA. 10"

FIGURE 3.6   (Continued)

3.5.2   FUEL POWER

For fuel-driven pumps (diesel, gasoline, propane or natural gas), Equation (3.8b) can be used to calculate the cost of operation. The energy cost for the pump in Example 3.1 is calculated as:

$$\text{Cost} = (\text{BHP}) * (T) * \left(\frac{1}{\text{gallon} / \text{BHP} - \text{hr}} \right) * (\$ / \text{gallon of fuel}), \qquad (3.8b)$$

In Example 3.1a, the BHP was calculated as 60.6 HP and the diesel fuel efficiency from Table 3.2. Is 13.75 BHP-hr/U.S. gal of fuel. Therefore, the cost of operation for 24 hours for a diesel engine-driven pump is:

$$\text{Cost} = (60.6) * (24) \left(\frac{1}{13.75} \right) * (\$1.25) = \$132.22 \text{ for 24 hours}$$

3.5.3   PUMP ECONOMICS

One of the pump design criteria is economics. The pump should be designed such that it has the lowest annual cost. Annual cost can be defined as annual capital cost + annual energy cost as shown below.

$$\text{Annual cost} = \text{CRF} (\text{capital cost}) + \text{Annual energy cost} \qquad (3.8c)$$

TABLE 3.2

Typical Performance Ratings for Engines

	Brake Horsepower Energy Rating	
Energy Source	BHP-hr/U. S. gal of fuel	BHP-hr/liter of fuel
Diesel	13.75	3.90
Gasoline	11.30	2.98
Propane	8.92	2.36

CRF is the capital recovery factor which can be defined based on life expectancy of the pump or loan period. The equation for CRF is defined as:

$$\text{CRF} = \frac{i(i+1)^n}{(1+i)^n - 1} \tag{3.8d}$$

in which i is the interest rate and n is the life expectancy of the pump + motor.

The following example shows how the economic decision is made.

TABLE 3.3

Formulas for Calculating Unknown Motor Condition

To Find	Single Phase	Three-Phase
Motor Current (Amperes)	$\dfrac{\text{Motor HP} \times 746}{E \times \text{Eff.} \times \text{P.F.}}$	$\dfrac{\text{Motor HP} \times 746}{1.73 \times E \times \text{Eff.} \times \text{P.F.}}$
	(Input)	(Input)
Motor Current (Amperes)	$\dfrac{\text{kW} \times 1000}{E \times \text{P.F.}}$	$\dfrac{\text{kW} \times 1000}{1.73 \times E \times \text{P.F.}}$
Motor Current (Amperes)	$\dfrac{\text{Kva} \times 1000}{E}$	$\dfrac{\text{kva} \times 1000}{1.73 \times E}$
Kilowatts (Input)	$\dfrac{I \times E \times \text{P.F.}}{1000}$	$\dfrac{I \times E \times 1.73 \times \text{P.F.}}{1000}$
Kva (Input)	$\dfrac{I \times E}{1000}$	$\dfrac{I \times E \times 1.73}{1000}$
Horsepower Nameplate Rating (Output)	$\dfrac{I \times E \times \text{Eff.} \times \text{P.F.}}{746}$	$\dfrac{I \times E \times 1.73\,\text{Eff.} \times \text{P.F.}}{746}$

E = voltage
I = amperage
P.F. = power factor, usually taken as 0.85
Kva = Kilovolt, number of revolutions per unit volt applied

TABLE 3.4
Average Efficiencies and Power Factors for Induction Motors, 3-Phase, 60-Cycle, 1800 RPM

	Efficiency			Power Factor		
H_p	½ Load	¾ Load	Full Load	½ Load	¾ Load	Full Load
10	83	85	85	65.5	76.5	82
25	86	87.5	87.5	70.5	79.5	83
50	91	91	91	77	83.5	87
75	91	91	91	80	86	87.5
100	91	92	92	86	90	90.5
150	91.5	92	92	83.5	89	90
200	92	93	93	83.5	88.5	90
250	92	93	93	82.5	87.5	89
300	90.5	92	92.5	82	87	89
300	91	92.5	92.7	83	89	90.5
500	92	93	93.5	87	87	89
600	92.5	93.5	93.8	87	90	92
700	92.5	93.5	93	87	89.5	92
800	92.5	93.5	93	87	89.5	90.5

Example 3.2

We want to select a pump for Q = 1600 gpm and TDH of 320. We have two possible candidates that can produce the desired Q and TDH. Which one is the better choice?

	Pump A	Pump B
Q	1600 gpm	1600 gpm
TDH	320 ft	320 ft
Combined Eff.	80%	76%
Capital Cost	$25,600	$20,500
Input Energy	120.56 kW	126.90
Life Expectancy	20 yrs	20 yrs

Interest rate, i = 7%

Hours of operation, 2190 hours

Electricity cost, \$0.15/kW h

$$CRF = \frac{0.07(1+0.07)^{20}}{(1+0.07)^{20}-1} = 0.0944$$

Input energy from Equations (3.4), (3.5) and (3.6) are:
Input energy (pump A) = 120.56 kW
Input energy (pump B) = 126.90 kW
$T = 2196$ hr

The total annual costs are:
Pump A = 0.0944*25,600 + 120.56*2190*\$0.15/kWh = \$42,020.6
Pump B = 0.0944*20,500 + 126.9*2190**0.15/kWh = \$43,621.85
Pump A is therefore a better choice.

Example 3.3:   Measuring Pump Efficiency

Pump efficiency can be measured in the field or in the laboratory by measuring input power at the electric panel and MHP using pumping rate and TDH of the pump.
The following example shows how to calculate efficiency of a pump powered by electric motor.
The pump is operating at

$Q = 900$ gpm
TDH = 140 ft

The TDH is calculated by summing the lift ($Z_1 - Z_2$) and friction loss (h_f) in the inlet pipe.
Amperage at the electric panel = 55 amp
Voltage: 460 V
Current and voltage are measured at the electric panel using suitable meters.
Using Table 3.3 the input power is calculated as:

$$kW\,(input) = \frac{I \times E \times 1.73 \times P.F}{1000} = \frac{55 \times 460 \times 1.73 \times 0.85}{1000} = 37.2 \text{ kW}$$

Using estimated motor efficiency of 91%, and assuming shaft efficiency of 100%,

$$MHP = \frac{37.2}{0.746} \times 0.91 = 45.38 HP$$

$$MHP = BHP = 45.38 = \frac{900 \times 140}{3960 \times Ef_p}$$

And from here, $Ef_p = 0.70$

3.6 PUMPS IN SERIES

Sometimes it is necessary to install more than one pump or impeller on a line to increase the flow or the pressure. When pumps are installed such that the water discharged from one pump goes directly through the next one, this is referred to as "pumps in series." In such case, the flow or Q is not increased, but the head or pressure is increased. Figure 3.7 illustrates two pumps in series. If each pump is capable of producing 100 feet of head, the combined head produced would be 200 feet. The combined curve on Figure 3.7 represents the head which pump A and B operating in series will generate at any Q.

The combined brake horsepower required by pumps in series is determined by adding the brake horsepower of each individual pump. It is common practice in turbine pump installations to "multistage." This is the practice of passing the flow in series through several impellers mounted on the same pump shaft. Each impeller or stage adds an incremental increase in pressure. The TDH versus discharge characteristic only shows the TDH for one stage. To find the total TDH output of pumps in series or of a multistage pump requires adding the TDH values of each stage or pump taken from the characteristic diagram.

Equations (3.9–3.11) show how the combined efficiency of multiple pumps installed in series is calculated.

Using conventional units, the combined efficiency is calculated as

$$E_{A+B} = \frac{Q\left(H_A + H_B\right)}{3960\left(\text{BHP}_A + \text{BHP}_B\right)} \qquad (3.9)$$

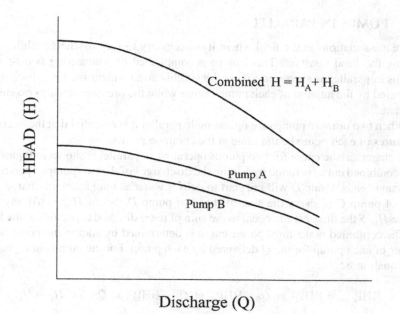

FIGURE 3.7   Head discharge curve for pumps operating in series.

where

    Q is in GPM
    H is in feet

and in the metric system where Q is in L/s and head is in m,

$$E_{A+B} = \frac{Q(H_A + H_B)}{76(\text{BHP}_A + \text{BHP}_B)} \qquad (3.10)$$

where

    Q is in liters/sec
    H is in meters

or

$$E_{A+B} = \frac{H_A + H_B}{\dfrac{H_A}{E_A} + \dfrac{H_B}{E_B}}, \qquad (3.11)$$

where E_{A+B} represents the combined efficiency.

In this example, one of impeller A and one of impeller B is installed in series. If more of each impeller are installed, then H_A or H_B become $n(H_A)$ or $n(H_B)$, where n represents the number of each impeller used.

3.7 PUMPS IN PARALLEL

There are situations in the field where it is necessary to vary discharge while maintaining the head constant. This can be accomplished by connecting two or more pumps in parallel as shown in Figure 3.8. In this arrangement the total discharge is increased by the amount of each pump output while the pressure remains essentially constant.

When two or more pumps are operating in parallel it is essential that the operating pressures of each pump be the same at the common points.

A characteristic curve for two pumps operating in parallel is shown in Figure 3.8. The combined curve is found by adding the discharges of the two pumps operating at the same heads. Pump D will not start to deliver water as long as the discharge pressure of pump C is above the shut-off head of pump D (below H_D, S). At any head above H_D, S the discharge is equal to the sum of the individual capacities at the head.

The combined brake horsepower curve is determined by adding the brake horsepower of each pump for the Q delivered by each pump. For the pumps in Figure 3.8 the equation is:

$$\text{BHP}_{C+D} = \text{BHP}_C \text{ at } Q_C + \text{BHP}_D \text{ at } Q_D + \text{BHP}_C \text{ at } Q_C \text{ for } H_C = H_D$$

The BHP is plotted against the combined flow.

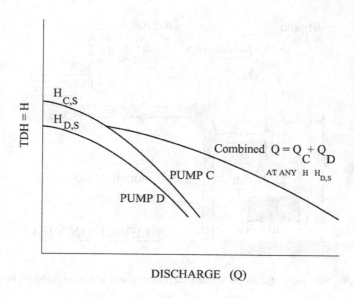

FIGURE 3.8   Head discharge curve for pumps operating in parallel.

The efficiency curve of the combined pumps can be determined by either of the following equations:

$$E_{C+D} = \frac{(Q_C + Q_D) \times \text{TDH}}{3960 \left(\text{BHP at } Q_C + \text{BHP at } Q_D \right)}$$

in which Q_C and Q_D are the discharges of pumps C and D in gpm, therefore;

$$E_{C+D} = \frac{Q_C + Q_D}{\dfrac{Q_C}{E_C} + \dfrac{Q_D}{E_D}} \tag{3.12}$$

where E_C and E_D are the efficiencies of pumps C and D at discharges Q_C and Q_D.

3.8   PUMP INTAKE AND DISCHARGE STRUCTURE

Intake structures are an integral part of pumping systems. The pump intakes need to be properly designed to avoid cavitation, air pumping, turbulence and vortex. Figures 3.9–3.11 demonstrate the proper intake structures for a typical centrifugal pump.

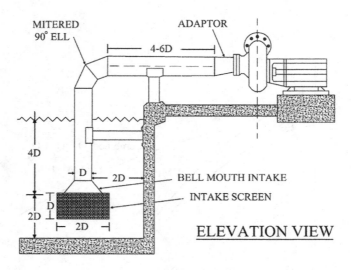

FIGURE 3.9 Side view of a properly installed intake structure for centrifugal pumps.

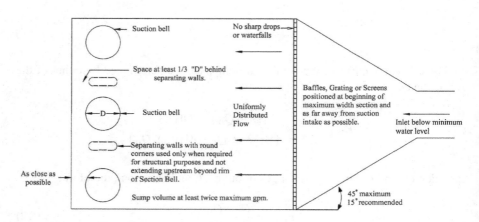

FIGURE 3.10 Top view of intake structure for multiple pumps.

3.9 OUTLET DESIGN

Figure 3.11 shows two pump discharge options for design of a free discharge pump outlet. The most common design, an open-air discharge (Figure 3.11b), results in a waste of energy as shown. Energy can be conserved by lowering the discharge point below the water surface (Figure 3.11a). However, it would be necessary to install an air vent in option (a) to avoid siphoning the water back into the source once the pump is turned off.

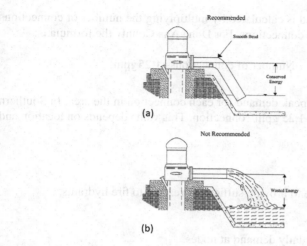

(a)

(b)

FIGURE 3.11 Pump discharge structure.

3.10 MUNICIPAL SYSTEM DESIGN

The following discussion describes the procedure for design of a municipal system that can satisfy the normal flow and pressure in addition to sufficient pressure for fire hydrants. The design can be accomplished using pipe network software such as KYPIPE or EPANET.

A municipal system should have the following characteristics:

1. System network pressure: 50–100 psi.
2. System should be able to provide the maximum daily demand.
3. Fire pressure should be at least 25 psi with two hydrants operating at 750 gpm each.
4. Pump should be able to deliver the maximum daily demand or more.
5. Reservoir should have sufficient capacity as described below.

3.10.1 SYSTEM NORMAL PRESSURE

a. Turn off the pump
b. Keep the maximum demand
c. Turn off fire hydrant
d. Calculate the pressure in each node

The pressure in nodes should be between 50 and 100 psi with the pump turned off.

3.10.2 PUMP CAPACITY

The system should be able to deliver maximum daily demand based on pump capacity.

Maximum daily demand is calculated by multiplying the number of connections by the peak demand in the connections. For Dona Ana County the formula is:

$$Q_{max} = \text{Number of connections} * 1.25 \, \text{gpm}$$

where 1.25 represents the peak demand for each connection in the area. In Southern New Mexico, this value is 1.25 gpm/connection. This value depends on location and climate conditions.

3.10.3 FIRE PRESSURE

The system should be able to provide sufficient pressure to fire hydrants.

   a. Turn off the pump.
   b. Keep the maximum daily demand at nodes.
   c. Set fire demand at 1500 gpm at critical nodes and check the pressure to make sure there is at least 25 psi at the fire node.

FIGURE 3.12 Municipal systems are designed for fire flow.

3.10.4 PUMP DESIGN

a. Pump capacity should be at least equal to maximum community demand, and preferably 10–20% higher.
b. Define the source of pumping as junction node, put a negative flow equal to pumping rate at the junction.
c. The resulting pressure at that junction is the TDH of the pump.
d. Design the pump and identify Q–H curve.
e. Change the source node to a reservoir, put a pump at the line between the source and the system, define the pump characteristics (at least three points), run the program and check the operation point.

3.10.5 RESERVOIR DESIGN

The capacity of the reservoir should be designed based on the following formula:

$$\text{Reservoir capacity (gallons)} = \text{max.demand} * 1440 + Q(\text{pump}) * 60 + 1500 * 120$$

Another approach is:

$$\text{Reservoir capacity} = \text{max.demand} * 4 * 1440,$$
$$\text{which equals four days of maximum demand.}$$

4 Deep-Well Turbine and Submersible Pump Curves

4.1 DEEP-WELL TURBINE PUMPS

Deep-well turbine pumps are a type of vertical shaft submersible pump. They are also known as vertical turbine pumps, vertical line-shaft pumps and line-shaft turbine pumps. They are borehole pumps whose pump shaft extends up to ground level, where it is coupled to a dry installed motor (e.g., hollow shaft motor) or gearbox (e.g., hollow shaft gearbox). They are multistage pumps and are usually fitted with mixed flow impellers. A typical deep-well turbine pump is shown in Figure 4.1.

4.2 SUBMERSIBLE PUMPS

A submersible pump, also called an electric submersible pump, is a pump that can be completely submerged in water. The motor is hermetically sealed and close coupled to the body of the pump.

A submersible pump thrusts water to the surface by transforming rotary energy into kinetic energy into pressure energy. This is done by the water being pulled into the pump, first in the intake, where the rotation of the impeller thrusts the water through the diffuser. From there, it goes to the surface.

The main benefit of a submersible pump is that it never has to be primed, because it is already submerged in the water. Submersible pumps are also very efficient since they expend less energy moving water into the pump. This is because the water pushes itself into the pump using the water pressure and therefore less energy has to be spent on moving water into the pump.

Also, while the pumps themselves aren't versatile, the selection certainly is. Some submersible pumps can easily manage solids, while some are only fit for liquids. Other benefits of this type of pump include low sound pollution and fewer issues associated with cavitation, since there is no "spike" in pressure as the water flows through the pump.

There are also some disadvantages with submersible pumps. The seals can become corroded with time. When that happens, water penetrates into the motor, rendering it

DOI: 10.1201/9781003287537-4

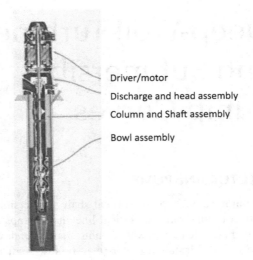

Driver/motor

Discharge and head assembly

Column and Shaft assembly

Bowl assembly

FIGURE 4.1 Vertical turbine pumps.

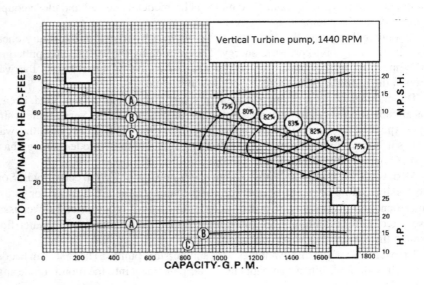

FIGURE 4.2 Vertical turbine pump 1440 RPM.

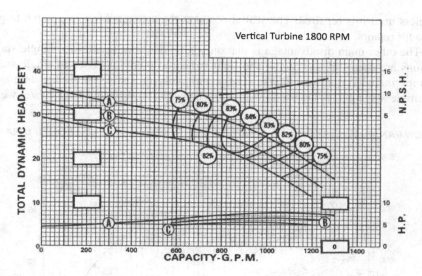

FIGURE 4.3   Vertical turbine pump 1440 RPM.

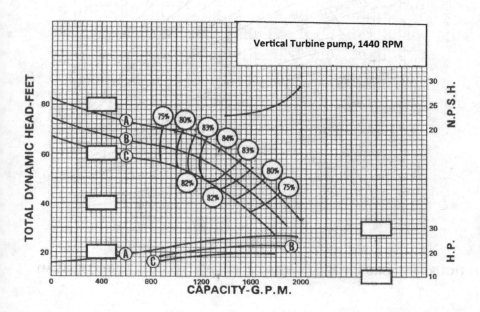

FIGURE 4.4   Vertical turbine pump 1440 RPM.

useless until it is repaired. The seal also makes the submersible pump tough to get into for repairs.

The other main disadvantage is that one pump does not fit all uses. Single-stage pumps are used for most home and light industrial pumping. Uses of single-stage pumps include aquarium filters, sewage pumping, or sump pumps for drainage. Multiple-stage pumps are used for anything beneath soil surface, such as water wells

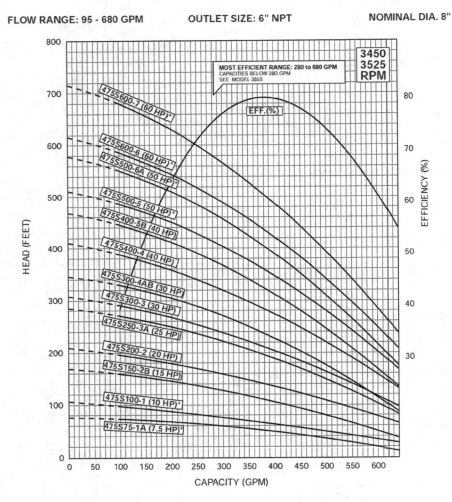

FIGURE 4.5   GRUNDFOS submersible curve model 445S (Speed: 3450 to 3525 RPM).

or oil wells. Also, pumps are made to work with low-viscous liquids like water, or viscous liquids like sewage.

Caution must be used with submersible pumps, which must be fully submerged. The water around a submersible pump actually helps to cool the motor. If it is used out of water, it can overheat (Figures 4.5–4.10).

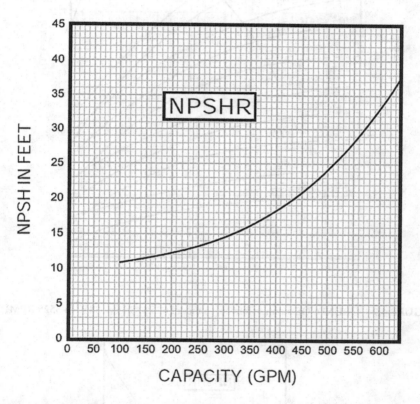

FIGURE 4.6 NPSHR curve for GRUNDFOS submersible curve model 445S (Speed: 3450 to 3525 RPM).

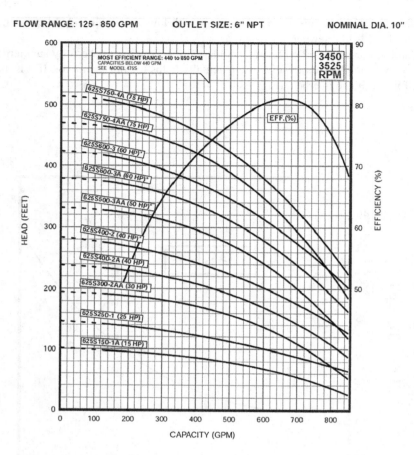

FIGURE 4.7 GRUNDFOS submersible curve model 625S (Speed: 3450 to 3525 RPM).

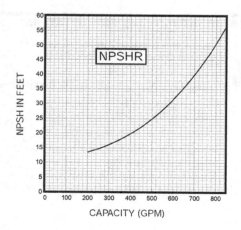

FIGURE 4.8 NPSHR for GRUNDFOS submersible curve model 625S (Speed: 3450 to 3525 RPM).

FLOW RANGE: 160 - 1100 GPM   OUTLET SIZE: 6" NPT   NOMINAL DIA. 10"

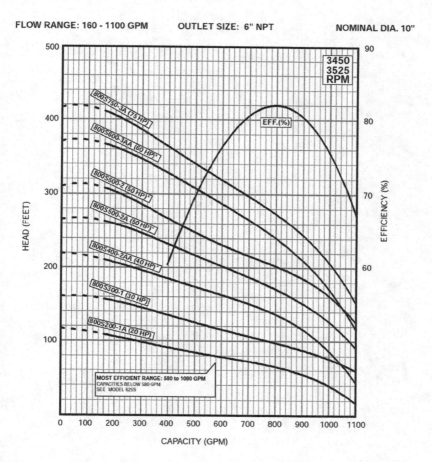

FIGURE 4.9   GRUNDFOS submersible curve model 800S (Speed: 3450 to 3525 RPM).

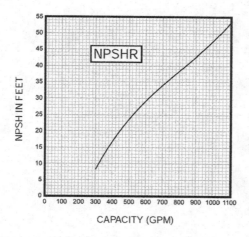

FIGURE 4.10   NPSHR for GRUNDFOS submersible curve model 800S (Capacity = 800 GPM; Speed: 3450 to 3525 RPM).

5 Open-Channel Hydraulics

5.1 SUMMARY EQUATIONS

Manning Equation:

$$Q = \frac{k}{n}(A)\left(R^{\frac{2}{3}}\right)S^{1/2}$$

$k = 1$ in SI system
$k = 1.486$ in English system

Trapezoidal channel dimensions:

DOI: 10.1201/9781003287537-5

Flow rate (Manning)
$$Q = \frac{k}{n}(A)\left(R^{\frac{2}{3}}\right)S^{1/2}$$

Cross-sectional area $A = by + my^2$
Top width $T = b + 2my$
Wetted perimeter $P = b + 2y\sqrt{m^2 + 1}$

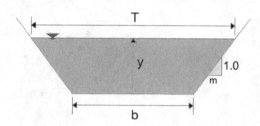

Trapezoidal Section

A = area of flow (L^2)
b = channel base width (L)
y = flow depth (L)
m = channel side slope
P = wetted perimeter (L)
T = water surface top width (L)
S= Slope of energy grade line =
bottom slope in uniform flow
R= A/P

Cross-section of a trapezoidal channel.
Circular Channel:

Circular Section

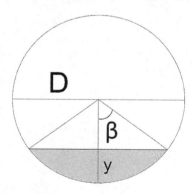

$$A = \frac{D^2}{4}(\beta - \sin\beta \cos\beta)$$
$$T = D\sin\beta$$
$$P = \beta D$$
$$\beta = \cos^{-1}(1 - \frac{2y}{D})$$

A = area of the flow
D = conduit or channel diameter (L)
Y = flow depth (L)
β = Angle between vertical line and
water surface (radians)
P = Wetted perimeter (L)
T = water surface top width (L)

Cross-section of a circular channel.

5.2 THE ENERGY PRINCIPLE IN OPEN CHANNELS

Water flowing in open channels obeys the law of conservation of energy. If the total hydraulic energy is represented by E, then energy at any point is

$$E = \text{Kinetic Energy} + \text{Pressure Energy} + \text{Elevation Energy}$$

or

$$E = \frac{V^2}{2g} + \frac{P}{\gamma} + y = \text{constant} \tag{5.1}$$

where:

V = average flow velocity (L/T)
g = acceleration of gravity (L/T^2)
y = depth of water (L)
γ = specific weight of water (F/L^3)

Equation (5.1) is the Bernoulli equation, which was first formulated by Leonhard Euler in its differential form. Based on the principle of conservation of energy, the total hydraulic energy decreases in the direction of flow (from section 1 to section 2), in an amount equal to the head loss h_L. The energy equation between sections (1) and (2) can be written as:

$$\frac{V_1^2}{2g} + \frac{P_1}{\gamma} + y_1 + z_1 = \frac{V_2^2}{2g} + \frac{P_2}{\gamma} + y_2 + z_2 + h_L \tag{5.2}$$

In open channel, Equation (5.2) takes the form

$$\frac{V_1^2}{2g} + y_1 + z_1 = \frac{V_2^2}{2g} + y_2 + z_2 + h_L \tag{5.3}$$

where y is the depth of water, Z is the vertical distance of the bottom of the channel from a horizontal reference line (base line). h_L is the portion of total hydraulic energy which is lost to friction. The sum $(Z + y)$ in Equation (5.3) is called the piezometric head, and represents the position of the hydraulic grade line at a given point. The difference between the total energy grade line and the hydraulic grade line is represented by the velocity head ($v^2/2g$) as shown in Figure 5.1.

There are two types of energy defined for open channel. These are total energy (Equation 5.1) and specific energy (Equation 5.4). Specific energy (E_s) is the energy in reference to the bottom of the channel at any point along the channel.

The specific energy at any point in open channels is defined as:

$$E_s = y + \frac{V^2}{2g} \tag{5.4}$$

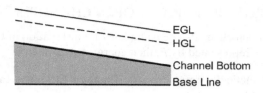

FIGURE 5.1   Diagram of energy grade line and hydraulic grade line along a channel.

The velocity in a given cross-section is variable. The velocity head as expressed in Equation 5.4 is a mean velocity, and not the true velocity. A channel section has a non-uniform velocity distribution and therefore the true velocity head is usually different than the velocity head computed using the mean velocity.

Coriolis introduced the concept of an energy coefficient α to correct the velocity head computed using the mean velocity. Experimental data suggest that α varies from 1.03 to 1.36 in straight prismatic channels (Chow, 1959).

Introducing the Coriolis coefficient into Equation 5.4 the expression becomes:

$$E_s = y + \frac{\alpha V^2}{2g} \tag{5.5}$$

Substituting the average velocity from the continuity equation ($Q = AV$) into Equation 5.5 gives an expression for energy in terms of flow rate and cross-sectional area of flow:

$$E_s = y + \frac{\alpha Q^2}{2gA^2} \tag{5.6}$$

Equation (5.6) will be used throughout this section to describe the energy status of flow in different hydraulic problems. For practical purposes, α will generally be assumed to be equal to one.

5.2.1 ALTERNATE FLOW DEPTHS

Let's take a rectangular channel section and use it to analyze the behavior of the specific energy E_s as a function of flow depth y. In a rectangular channel section, the flow rate per unit width of channel is represented by $q = Q/b$, where b is the channel base width and Q the total flow rate. Substituting a unit flow rate into Equation (5.6) and assuming $\alpha = 1$, results in:

$$E_s = y + \frac{V^2}{2gy^2} = y + \frac{Q^2}{2gA^2}$$

Or

$$\left(E_s - y\right)y^2 = \frac{q^2}{2g} = \text{constant} \tag{5.7}$$

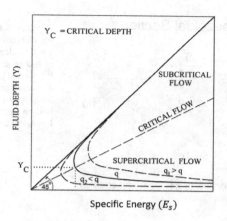

FIGURE 5.2   Generalized specific energy curve.

where q is the flow rate per unit width of the rectangular channel

The curve has asymptotes given by $(E_s - y) = 0$ and $y = 0$, and has the shape shown in Figure 5.2.

The plot of E_s vs. y is called the specific energy curve. This curve defines two flow regimes, one of deeper and slower flow, and the other of shallower and faster flow. These flow regimes will be treated in greater depth in the following section, as well as the concepts of critical energy (E_c) and critical depth (y_c).

For sections other than rectangular, the total flow rate and area of flow as a function of depth must be used with Equation (5.6).

Two types of cross-section will be explored in this chapter: trapezoidal and circular. Triangular and rectangular channels are subsets of trapezoidal channel.

5.3   TRAPEZOIDAL CHANNEL

Open channels come with various cross-sectional characteristics that include circular, trapezoidal, triangular, rectangular and irregular. Trapezoidal sections are the most common type of open channel, but the other types of cross-sections are also encountered. The most hydraulically efficient open channel is a semi-circular channel. A semi-circular channel has the smallest wetted perimeter, thus has the highest flow rate per unit of cross-sectional area. However, semi-circular channels are structurally unstable. Therefore, trapezoidal channels which are closest to a semi-circle are considered most efficient. Among the trapezoidal channels the one with side slope of 60 degrees is closest to a semi-circle. In practice, however, the 60-degree angle is too steep and quite often open channels are built with a 45-degree side slope.

The equations of flow in open channel are based on Manning equation. The Manning equation is described as:

$$Q = \frac{k}{n}(A)\left(R^{\frac{2}{3}}\right)S^{1/2} \tag{5.8}$$

Trapezoidal Section

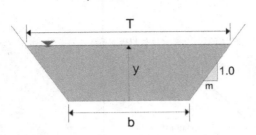

A = area of flow (L^2)
b = channel base width (L)
y = flow depth (L)
m = channel side slope
P = wetted perimeter (L)
T = water surface top width (L)
S= Slope of energy grade line = bottom slope in uniform flow
R= A/P

FIGURE 5.3   Cross section of a trapezoidal channel.

where

Q = Flow rate
n = Manning roughness
A = Cross-sectional area
R = Hydraulic radius = A/P
S = Longitudinal slope
k = 1 in SI system and 1.486 in English system

The parameters in Equation (5.8) for a trapezoidal channel are described below.

Cross-sectional area	$A = by + my^2$
Top width	$T = b + 2my$
Wetted perimeter	$P = b + 2y\sqrt{m^2 + 1}$

Circular Section

$$A = \frac{D^2}{4}(\beta - \sin \beta \cos \beta)$$
$$T = D \sin \beta$$
$$P = \beta D$$
$$\beta = cos^{-1}(1 - \frac{2y}{D})$$

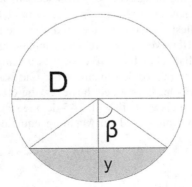

A = area of the flow
D = conduit or channel diameter (L)
Y = flow depth (L)
β = Angle between vertical line and water surface (radians)
P = Wetted perimeter (L)
T = water surface top width (L)

FIGURE 5.4   Cross-section of a circular channel.

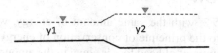

FIGURE 5.5 Effect of elevation change on water profile.

5.4 CIRCULAR CHANNEL

The dimensions of a circular channel are calculated as follows:

5.4.1 APPLICATIONS OF THE ENERGY EQUATION IN OPEN CHANNEL

Example 5.1

A rectangular channel (Figure 5.5) with bottom width of 4 ft, flow depth of 10 ft, containing a flow rate of Q = 100 CFS smoothly drops its bottom elevation by 2 ft. Determine the new depth.

Solution:

A smooth change in elevation will have negligible head loss. Therefore, between sections 1 and 2 energy will be preserved so that $E_1 = E_2$. Using the energy equation:

$$E_1 = y_1 + \frac{Q^2}{2g A_1^2} = 10 + \frac{100^2}{2g(4\times10)^2} + 2 = 12.1 \text{ ft}$$

Assuming no head loss,

$$E_1 = E_2 = 12.1 = y_2 + \frac{100^2}{2g(4y_2)^2}$$

Which results in

$$y^3 - 12.1y^2 + 9.7 = 0$$

And by trial and error, y_2 = 12.03 ft
What will happen if the bottom elevation rises 2 ft?
Using the same process,

$$E_1 = E_2 + 2$$

$$E_2 = 10.1 - 2 = 8.1$$

Which results in

$$y^3 - 8.1y^2 + 9.7 = 0$$

And by trial and error

$$y_2 = 7.95 \text{ ft}$$

5.5 CRITICAL FLOW

As shown in Figure 5.2, with the same amount of energy, water can assume two possible depths. Due to the principle of conservation of energy, water may adjust its velocity such that the amount of head loss is equal to the amount of available gravitational energy (Garcia, 1986). Thus water will move slowly when the channel slope is mild and will move faster when the channel slope is steep. Water adjusts its velocity to match the available energy. The depth of the water when the slope of energy grade line (head loss per unit of length) is equal to the slope of the bottom of the channel is called normal depth. When the slope is mild, water is typically flowing at sub-critical condition and when the slope is steep the water typically flows at super-critical condition. The deflection point in Figure 5.2 when the water moves from sub-critical to super-critical flow is called critical depth. Figure 5.6 shows a case where water transitions from sub-critical to super-critical condition (Garcia, 1986). In order for water to transition from sub-critical to super-critical flow, it has to pass through the critical flow as shown in Figure 5.2.

The critical depth is calculated by setting the derivative of the specific energy equation to zero, as shown below.

In order to obtain an expression for the Froude number in terms of channel flow rate and cross-sectional area, the energy equation must be differentiated with respect to flow depth y and made equal to zero for finding the minimum.

$$E_s = y + \frac{V^2}{2g}$$

$$E_s = y + \frac{Q^2 A^{-2}}{2g}$$

$$\frac{dE_s}{dy} = 1 + \left[\frac{Q^2}{2g}\left(-2A^{-3}\right)\right] \times \left(\frac{dA}{dy}\right) = 0$$

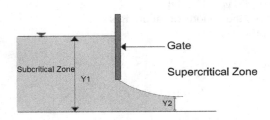

FIGURE 5.6   Sub-critical and super-critical flow across a sluice gate.

$$\frac{dA}{dy} = T \tag{5.9}$$

$$\frac{dE_s}{dy} = 1 - \frac{Q^2 T}{gA^3} = 0, \tag{5.10}$$

and

$$\frac{Q^2 T}{gA^3} = 1 = \text{Fr}^2, \tag{5.11}$$

Equation (5.11) is the equation of critical flow and the "Fr" stands for Froude number. The Froude number can be written as a function of velocity:

$$\text{Fr}^2 = \frac{V^2 T}{gA} \tag{5.12}$$

For rectangular channel, the Froude number can be simplified to:

$$\text{Fr} = \frac{V}{\sqrt{gy}} \tag{5.13}$$

Critical flow conditions exist when water in a channel moves as fast relative to the banks as a wave resulting from a small disturbance moves relative to the water (Henderson, 1966). The wave velocity with which a disturbance in open-channel flows tends to move over the water surface is given by $c = \sqrt{gy}$

When the stream velocity is less than the critical velocity, the wave caused by a disturbance travels upstream from the view of a stationary observer. This condition is called a sub-critical flow and $v < \sqrt{gy}$ Conversely, when the stream velocity is greater than the critical velocity the wave front from a small disturbance travels downstream. This condition is called super-critical flow and $v > \sqrt{gy}$.

The Froude number will represent the parameter, differentiating sub-critical, critical and super-critical flow thus:

Fr < 1 Sub-critical

Fr = 1 Critical

Fr > 1 Super-critical

Since in sub-critical flow the wave from a small disturbance moves upstream, a control mechanism such as a sluice gate can influence the flow upstream. Therefore, sub-critical flow is subject to downstream control. On the other hand, in super-critical flow a wave disturbance moves downstream and therefore cannot be subject to downstream control (Henderson, 1966).

5.6   APPLICATIONS OF CRITICAL FLOW

As is shown in Equation (5.11), at critical flow the discharge rate is only dependent on the channel geometry, thus by measuring or calculating critical depth, the discharge rate can be calculated. This concept is the foundation of flow measuring devices such as Parshall flume, RBC flume and various types of weirs.

The following example illustrates this application.

Example 5.2

A canal is to convey 1.2 m³/s. Find the critical depth if:

  (a) The section is trapezoidal; $b = 1.5$ m and $m = 2$.
  (b) The section is circular; $D = 1.0$ m.
  (c) The section is rectangular; $B = 2.0$ m.

Solution:

The solution to the problem requires the use of some trial procedure, since the critical depth relationship is implicit in the critical depth y_c. From Equation (5.11) we can write

$$Q^2 T = gA^3$$

$$\text{Let } f(y_c) = 0$$

and

$$f(y_c) = Q^2 T - gA^3 = 0$$

  (a) For a trapezoidal channel, the area (A) and top width (T) must be replaced into the function $f(y_c)$.

$$f(y_c) = Q^2 (b + 2my_c) - g(by_c + my_c^2)^3 = 0$$

The only unknown on the non-linear function above is the depth (y) which can be solved iteratively. and Yc = 0.308.

  (b) Similarly, the area and top width for a circular channel must replace T and A in the above equation for circular pipe:

$$f(y_c) = Q^2 [D \sin \beta] - g\left[\frac{D^2}{4}(\beta - \sin \beta \cos \beta)\right]^3 = 0$$

where

$$\beta = \cos^{-1}\left(1 - \frac{2y_c}{D}\right)$$

The function can be solved for critical depth iteratively. and Yc = 0.63.

(c) Finally, the same procedure can be used with a rectangular section which results in a linear function:

$$f(y_c) = Q^2T - g[2y_c]^3 = 0$$

$$(1.2)^2(2) - g(2y_c)^3 = 0$$

which results in

$$Y_c = 0.34\,\text{m}$$

5.7 THE MOMENTUM PRINCIPLE IN OPEN CHANNELS

The derivation of the hydraulic jump equation is based on the momentum principle. The momentum function can be derived from the analysis of a free body diagram applied to a body of fluid as shown in Figure 5.7 (Garcia, 1986). The summation of forces in the X and Y directions are:

$$\Sigma F_X = \rho Q(V_{2X} - V_{1X}), \qquad (5.14)$$

And in y direction are

$$\Sigma F_Y = \rho Q(V_{2Y} - V_{1Y}), \qquad (5.15)$$

Since the average velocity at sections (1) and (2) can be represented by the ratio of flow rate to area of flow, Equation (5.14) becomes:

$$\Sigma F_X = \rho Q\left(\frac{Q}{A_2} - \frac{Q}{A_1}\right) = \rho Q^2\left(\frac{1}{A_2} - \frac{1}{A_1}\right) \qquad (5.16)$$

The summation of forces in the y direction, as well as the friction component f, will be ignored in this analysis. Forces in the vertical direction are negligible since the slope is very small. Applying Equation (5.16) to the free body diagram in Figure 5.7 yields:

$$\Sigma F_X = \rho Q^2\left(\frac{1}{A_2} - \frac{1}{A_1}\right), \qquad (5.17)$$

FIGURE 5.7 Forces acting on a hydraulic jump.

The pressure forces are given by:

$$F_{P_1} = \gamma \bar{y}_1 A_1$$

and

$$F_{P2} = \gamma \bar{y}_2 A_2$$

where $\bar{y}_1$ and $\bar{y}_2$ are the distances from the water surface to the centroid of the area where the pressure force is acting, as shown in Figure 5.8.

$$\gamma \bar{y}_1 A_1 - \gamma \bar{y}_2 A_2 = \frac{\gamma}{g} Q^2 \left(\frac{1}{A_2} - \frac{1}{A_1} \right)$$

$$\gamma \bar{y}_1 A_1 - \gamma \bar{y}_2 A_2 = \frac{Q^2}{g} \left(\frac{1}{A_2} - \frac{1}{A_1} \right)$$

Combining the two functions results in Equation (5.18):

$$\bar{y}_1 A_1 + \frac{Q^2}{gA_1} = \bar{y}_2 A_2 + \frac{Q^2}{gA_2} \tag{5.18}$$

From Equation (5.18) and knowing that momentum is the application of a force per unit weight of fluid ($M = F/Y$), the momentum function becomes:

$$M = A\bar{y} + \frac{Q^2}{gA}, \tag{5.19}$$

where:

M = Momentum
$A\bar{Y}$ = First moment of area (L^3)

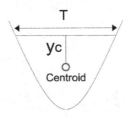

FIGURE 5.8   Shows the distance from water surface to centroid of an area.

Q = Flow rate (L³/T)
g = Acceleration of gravity (L/T²)
A = Cross-sectional area of flow (L²)

The first moment of area $A\bar{Y}$ must be determined for each channel section. The general relationship is derived from taking moments about the free water surface.

$$d\left(A\bar{y}\right) = d\left[A\left(\bar{y} + dy\right) + T\left(dy\right)\left(\frac{dy}{2}\right) - \bar{y}\left(A\right)\right] \qquad (5.20)$$

Considering that the second-order differential $(dy)^2$ is small, compared to the other terms, the sum of moments becomes:

$$d\left(A\bar{y}\right) = d\left(A\bar{y} + A.dy - A\bar{y}\right)$$

For a change dy in the depth of water, the change in the static moment of water area ($A\bar{y}$) is then given by the relationship:

$$d\left(A\bar{y}\right) = A.dy \qquad (5.21)$$

In a trapezoidal section, the area of flow is composed of two geometric shapes, a rectangle and a triangle. The differential change in the static moment of area is given by:

$$d\left(A\bar{y}\right) = by.dy + 2\left(\frac{my^2}{2}\right)dy$$

$$d\left(A\bar{y}\right) = \left[by + my^2\right]dy$$

Integrating both sides from zero to the depth (y) yields the moment for trapezoidal channel:

$$A\bar{y} = \frac{by^2}{2} + \frac{my^3}{3} \qquad (5.22)$$

Therefore, the momentum function for a trapezoidal cross-section would be

$$M = A\bar{y} + \frac{Q^2}{gA} = \frac{by^2}{2} + \frac{my^3}{3} + \frac{Q^2}{gA} \qquad (5.23)$$

For a rectangular channel, $m = 0.0$ in Equation (5.23) and for a triangular channel, $b = 0.0$.

The moment of area for a circular section is found in a similar way. The differential change is now given by:

$$d\left(A\bar{y}\right) = A.dy$$

$$d(A\bar{y}) = \frac{D^2}{4}(\beta - \sin\beta\cos\beta)dy$$

Since

$$y = \frac{D}{2}(1 - \cos\beta)$$

$$dy = \frac{D}{2}\sin\beta\,d\beta$$

And

$$A.dy = \frac{D^3}{5}(\beta\sin\beta - \sin^2\beta\cos\beta)d\beta$$

Integration of the first term yields:

$$\int_0^\beta \beta\sin\beta d\beta = \sin\beta - \beta\cos\beta$$

And the second term yields:

$$\int_0^\beta \sin^2\beta\cos\beta d\beta = \frac{\sin^3\beta}{3}$$

And

$$A\bar{y} = \frac{D^3}{5}\left(\sin\beta - \beta\cos\beta - \frac{\sin^3\beta}{3}\right) \tag{5.24}$$

Therefore, the Momentum for a circular section becomes,

$$M = \frac{D^3}{5}\left(\sin\beta - \beta\cos\beta - \frac{\sin^3\beta}{3}\right) + \frac{Q^2}{gA} \tag{5.25}$$

5.8 APPLICATION OF THE MOMENTUM PRINCIPLE IN A HYDRAULIC JUMP

A very common application of the momentum function is the occurrence of a hydraulic jump when flow changes from super-critical to sub-critical, energy is dissipated, but momentum is conserved.

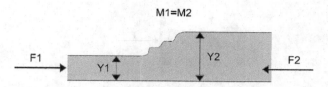

FIGURE 5.9   Schematic representation of a hydraulic jump.

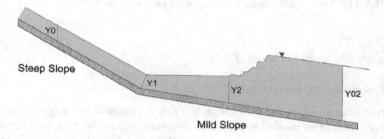

FIGURE 5.10   Abrupt change in channel slope causes a hydraulic jump to occur.

Knowing one of the conjugate depths allows the calculation of the second depth by simply making the momentum at section 1 (M_1) equal to the momentum at section 2 (M_2):

$$M_1 = M_2$$

Which results in

$$A\bar{y}_1 + \frac{Q^2}{gA_1} = A\bar{y}_2 = \frac{Q^2}{gA_2}$$

(5.26)

The equations for the first moment of area and cross-sectional area of flow for each specific channel section are used in Equation (5.26) to solve a variety of problems. When there is an abrupt change in channel slope, a hydraulic jump is likely to form. Analysis of the problem requires that the location of the hydraulic jump be determined. It can either occur in the steep slope or the mild slope in the channel.

A simple method to determine where the hydraulic jump occurs is to solve for MO_1 and MO_2.

If $MO_1 < MO_2$, the hydraulic jump occurs in the steep section of the channel. Conversely, if $MO_1 > MO_2$, the jump occurs in the mild section of the channel (Garcia, 1986).

5.9   CHANGE IN MOMENTUM

A change in momentum across a hydraulic structure originates a force against the structure, per unit weight of the fluid changing momentum. In mathematical form:

$$F = \gamma\left(\Delta M\right)$$

(5.27)

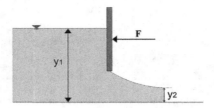

FIGURE 5.11   Momentum change across a sluice gate, $E_1 = E_2$, $M_1 \# M_2$.

Where,

   γ: is specific weight of water (F/L^3)
   and
   ΔM: is change in momentum (L^3)

A typical example is the flow through a sluice gate, where energy is essentially conserved (except for minor losses), but momentum drastically changes. The momentum change exerts a force against the gate (Figure 5.11) which can be calculated using Equation 5.27.

Example 5.3: Hydraulic Jump

Water flows in a trapezoidal channel with $b = 6.0$ m, side slope $m = 2$ carrying a flow rate of 28 m³/s. If the downstream depth of a hydraulic jump is 2.44 m, find the upstream depth.

Solution:

This problem requires that the value of the momentum function be calculated, using the downstream depth of 2.44 m.

$$M_1 = \left[\frac{by^2}{2} + \frac{my^3}{3} \right] + \frac{Q^2}{g\left[by + my^2 \right]} \tag{5.23}$$

$$M_2 = 27.55 + 3.01014 = 30.56 \text{ m}^3$$

Once M is known, then the equality $M_1 = M_2$ must be applied between the upstream and downstream sections in order to find the upstream depth y_1. The equation to be solved is implicit in y and must be solved iteratively. The equation to be solved has the form:

$$\left[\frac{by_1^2}{2} + \frac{my_1^3}{3} \right] + \frac{Q^2}{g\left[by_1 + my_1^2 \right]} - M_2 = 0$$

The non-linear function above can be solved by trial and error which results in:

$$Y_1 = 0.391 \text{m}$$

$$A_1 = 2.64 \text{m}^2$$

Rectangular Channel

A simple case of momentum application can be shown in a rectangular channel ($m = 0$) and bottom width of 1. In this case, Equation (5.23) for a unit bottom width becomes:

$$M = \frac{y^2}{2} + \frac{V^2 \cdot y}{g}$$

and the equation of hydraulic jumps becomes:

$$\frac{y2}{y1} = \frac{1}{2}\left[-1 + \sqrt{\left(1 + 8\left(Fr1^2\right)\right)}\right]$$

Example 5.4

A rectangular channel with bottom width of 6 m is carrying 25 m³/s at 1 meter of depth. The channel goes through a hydraulic jump. Determine the downstream depth and the head loss through the hydraulic jump.

From the momentum equation for a rectangular channel we have

$$\frac{y2}{y1} = \frac{1}{2}\left[-1 + \sqrt{\left(1 + 8\left(Fr1^2\right)\right)}\right]$$

$$V1 = \frac{25}{1 \times 6} = 4.17\,\frac{m}{s}$$

$$Fr1^2 = \frac{V1^2}{gy1} = 1.77$$

The flow is super-critical, and,

$$\frac{y2}{1} = \frac{1}{2}\left[-1 + \sqrt{\left(1 + 8\left(Fr1^2\right)\right)}\right],$$

$$y_2 = 1.447\ m$$

The head loss due to this hydraulic jump is calculated as

$$\text{Head loss} = dE = E_1 - E_2 = 1.886 - 1.87 = 0.017\,m$$

5.10   NON-UNIFORM FLOW (GRADUALLY VARIED FLOW)

A flow where the depth gradually varies along the length of the channel will be considered in this section. Two conditions that must prevail in this analysis approach are (Chow, 1959):

(1) The flow is steady, that is, there is no change in flow properties with time.
(2) The streamlines are essentially parallel to ensure hydrostatic pressure distribution.

The basic assumption behind the gradually varied flow analysis is that the head loss at a given channel section is the same as for uniform flow. This implies that the energy slope in the gradually varied flow situation can be estimated from uniform flow formulas.

Taking all these factors into consideration, the gradually varied flow equation for prismatic channels can be derived from the energy equation as follows:

$$E = y + \frac{V^2}{2g} + z$$

$$\frac{dE}{dx} = \frac{d}{dx}\left(y + \frac{V^2}{2g}\right) + \frac{dZ}{dx}$$

Substituting $\frac{dE}{dx} = -S_f$ and $\frac{dz}{dx} = -S_0$, the equation becomes

$$\frac{dE_s}{dX} = S_0 - S_f, \tag{5.28}$$

Or

$$\frac{E_{s2} - E_{s1}}{S_0 - S_f} = dX \tag{5.29}$$

Or

$$\frac{E_{s1} - E_{s2}}{S_f - S_0} = dX \tag{5.30}$$

From Equation (5.10) we have that:

$$\frac{dE}{dX} = 1 - Fr^2$$

By dividing Equation (5.30) by the expression of dE/dy, the gradually varied flow equation in terms of flow depth becomes:

$$\frac{dy}{dx} = \frac{S_0 - S_f}{1 - Fr^2}, \tag{5.31}$$

Figure 5.12 shows the profile of a gradually varied flow where depth upstream of free fall is calculated using the step method.

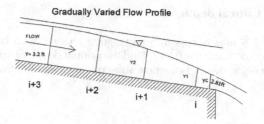

FIGURE 5.12   Water-surface profile in gradually varied flow for Example 5.6.

5.11   NORMAL DEPTH

The Manning equation is the closed-form solution of the conservation of energy equation (Equation 5.1). Normal depth Y_n is the depth at which uniform flow will occur in an open channel. Normal depth may be determined by writing the Manning equation for discharge:

English units

$$Q = \frac{1.49}{n} AR^{2/3}S^{1/2}$$

(from equation 5.8)

In SI Units

$$Q = \frac{1}{n} A.R^{2/3}S^{1/2}$$

(from equation 5.8)

and substituting for A and R expressions involving y and other necessary dimensions of the channel cross-sections. The resulting equation requires a trial-and-error-solution or normal depth may be computed through the use of tables.

In the Manning equation, if S is used as the bottom slope of the channel, then the calculated depth is normal depth, but if the slope of energy grade line is used as S, then the calculated depth is not necessarily equal to normal depth.

5.12   CRITICAL DEPTH

The critical depth y_c for flow in an open channel is defined as that depth for which the specific energy (sum of depth and velocity head) is a minimum. It can show mathematically that critical depth occurs in a channel when

$$\frac{Q^2}{g} = \frac{A^3}{T},$$

(From Equation 5.11)

where T is the surface width. On a mild slope uniform flow is sub-critical, while on a steep slope it is super-critical.

Example 5.5: Critical depth

A discharge of 3.5 m³/s occurs in a rectangular channel with bottom width of 1.53 m and $S = 0.002$ and $n = 0.012$. Find the normal depth for uniform flow and determine the critical depth. Is the flow sub-critical or super-critical?

Solution:

$$Q = AR^{2/3}S^{1/2} = Q = \frac{1}{0.012}1.53y_n\left(\frac{1.53y_n}{1.53+2y_n}\right)^{2/3}0.002^{1/2} = 3.5 \text{ m}^3/s$$

From the above equation, $y_n = 1.060$ m. Critical depth occurs when

$$\frac{Q^2}{g} = \frac{A^3}{T} \text{ or } \frac{3.5^2}{9.81} = \frac{(1.53y_c)^3}{1.53}$$

for which $y_c = 0.81$ m. Since $y_n > y_c$ the flow is subcritical.

5.13   FREE FALL

If the flow in a channel is sub-critical, critical depth should theoretically occur at a free fall (Figure 5.12) since the specific energy is at a minimum at that point. However, curvature of the streamlines in the vicinity of the fall alters the flow conditions and results in a depth at the brink less than y_c. The depth at a free fall has been observed experimentally to be about $0.7y_c$ for sub-critical flow, and under such conditions critical depth occurs at about $4y_c$ upstream from the brink. In contrast, if the flow is super-critical, the depth at the brink will be only slightly less than the normal depth, which is necessarily below the critical depth.

Example 5.6: Gradually varied flow

A discharge of 160 cfs occurs in a rectangular open channel 6 ft wide with $S_0 = 0.002$ and $n = 0.012$. If the channel ends in a free outfall as shown in Figure 5.12 calculate the depth at the brink, y_n and y_c. Determine the shape of the water-surface profile upstream of the brink.

Solution:

$$Q = \frac{1.49}{n}AR^{2/3}S^{1/2} = \frac{1.49}{n} \times 6y_n \times \left(\frac{6y_n}{6+2y_n}\right)^{2/3} \times 0.002^{1/2} = 160\text{cfs}$$

By trial and error, $y_n = 3.5$ ft. Critical depth occurs when

$$\frac{Q^2}{g} = \frac{A^3}{T} \text{ or } \frac{160^2}{32.2} = \frac{(6y_c)^3}{6}$$

which gives $y_c = 2.81$ ft. Since $y_n > y_c$ the flow is sub-critical and the water-surface profile is M_2. The depth at the outfall is approximately $0.7y_c = 2.0$ ft. Critical depth occurs at about $4y_c = 11$ ft upstream from the brink. Computations for the water-surface profile using Equation 5.30 are shown in Table 5.1.

5.14 CALCULATION OF FLOW DIMENSIONS IN A CIRCULAR CHANNEL

Due to the nonlinear nature of circular pipe equations, a simplified graphical approach has been proposed to solve the Manning equation for a partially full circular pipe. Figure 5.14 is a dimensionless curve for a circular channel where flow dimensions can be calculated as a function of the flow of a full pipe.

Example 5.7

A circular storm sewer with diameter of 4 ft, bottom slope of 0.001, and roughness of 0.013 is flowing full.

1. What is the flow?
2. What is the velocity

Solution:

$$Q = 1/n(A)(R^{2/3})S^{0.5} = (1/0.013)*(3.14*4.0)*(4/4)^{2/3}*(0.001^{0.5}) = 30.55\,\text{cfs}$$

$$V = Q/A = 30.55/12.56 = 2.43\,\text{ft/s}$$

If the flow depth is only 3 ft, what would be the flow and velocity?

$$d/D = 3/4 = 0.75$$

Using a ratio of 0.75 in the y axis Figure 5.14 results in Q/Q_{full} of 0.95.

$$Q = 30.55(0.95) = 29\,\text{cfs},$$

$V_{full} = 2.43$ ft/s, and
$V/V -$ Full $= 1.15$
$V = 1.15 \times 2.43 = 2.8$ ft/s

5.15 DESIGN CRITERIA FOR OPEN CHANNEL

The design of an open channel should be based on the following criteria:

1. It should carry the desired flow capacity.
2. It should have non-erosive and non-depositing velocity.
3. It should have stable structure (e.g. proper side slope).
4. It should be ethical (economical, safe, sustainable etc.).
5. It should have proper freeboard.

TABLE 5.1

Computation of Water-Surface Profile for Example 5.5

y, ft	A, ft²	$B + 2y$	R	V	$V^2/2g$	$Y + V^2/2g$	$\Delta(y + V^2/2g)$	V (Average)	R (Average)	S_f	$S_f - S_0$	dx	Σx
2.81	16.86	11.62	1.451	9.49	1.398	4.208							
							0.005	9.34	1.463	0.00341	0.00141	3.55	3.55
2.9	17.4	11.8	1.475	9.2	1.313	4.213							
							0.014	9.04	1.487	0.00312	0.00112	12.5	16
3	18	12	1.5	8.89	1.227	4.227							
							0.022	8.75	1.512	0.00286	0.00084	26	42
3.1	18.6	12.2	1.525	8.6	1.149	4.249							
							0.029	8.47	1.536	0.00262	0.00062	47	89
3.2	19.2	12.4	1.548	8.33	1.078	4.278							

TABLE 5.2
Values of the Roughness Coefficient *n* for Various Materials

Surface Material	Manning's Roughness Coefficient
Asphalt	0.016
Brickwork	0.015
Cast iron, new	0.012
Clay tile	0.014
Concrete – steel forms	0.011
Concrete (cement) – finished	0.012
Concrete – wooden forms	0.015
Concrete – centrifugally spun	0.013
Earth	0.025
Earth channel – clean	0.022
Earth channel – gravelly	0.025
Earth channel – weedy	0.030
Earth channel – stony, cobbles	0.035
Floodpains – pasture, farmland	0.035
Floodpains – light brush	0.050
Floodpains – heavy brush	0.075

FIGURE 5.13 Large unlined delivery canal (West Side Canal, Las Cruces, NM) used for irrigation.

5.16   RECOMMENDED SIDE SLOPE

The recommended side slope for a lined channel depends on economic factors. The most hydraulically efficient trapezoidal cross-section is a side slope of 60 degrees ($m = 0.577$). For in-situ construction of concrete lined channels, the most practical side slope is 1/1, which is ($m = 1$) since wet concrete will not be stable at higher slopes.

5.17   RECOMMENDED VELOCITIES

Table 5.3 shows recommended maximum velocities for various materials. For unlined channels, it is recommended to maintain velocities between 0.6–0.8 m/s. The lower velocity is to avoid sedimentation and the higher value is to prevent erosion. For more stable material such as clay or gravel, a velocity higher than 0.8 m/s may be used as is shown in Table 5.3. For concrete or rock lined channels a velocity range of 0.6–6 m/s may be used. If the velocity is too high and erosive for an unlined channel, the velocity can be reduced by reducing the slope which may necessitate construction of drop structures.

TABLE 5.3
Maximum Permissible Velocities (V) Recommended for Various Soil Conditions

Material	V (m/s)	n
Fine sand	0.5	0.020
Vertical sandy loam	0.58	0.020
Silt loam	0.67	0.020
Firm loam	0.83	0.020
Stiff clay	1.25	0.025
Fine gravel	0.83	0.020
Coarse gravel	1.33	0.025
Gravel	1.2	–
Disintegrated rock	1.5	–
Hard rock	4.0	–
Brick masonry with cement pointing	2.5	–
Brick masonry with cement plaster	4.0	–
Concrete	6.0	–

TABLE 5.4
Recommended Side Slope for Earthen and Lined Channel

Material	Slope, m:1
Peat soil	1.5
Clay	1.5
Loose soil	2–3
Concrete	1 (practical), 0.577 (theoretical)

TABLE 5.5
Recommended Freeboard for Open Channel

Discharge, m³/s	Freeboard, m
1–5	0.5
5–10	0.6
10–30	0.75
30–150	0.9

5.18 FREEBOARD

Freeboard is the distance from the maximum depth of water in a channel to the height of the embankment. Freeboard depends on various parameters, and may be regulated by agencies overseeing the safety of the canal. One simple rule of thumb is:

$$\text{Freeboard} = 1/6 \text{ of depth.}$$

More specific guidelines are shown in Table 5.5.

Example 5.8: Design of trapezoidal channel

The following example shows how an unlined trapezoidal channel can be designed for a given flow:

$$\text{Planned } Q = 1\text{m}^3/\text{s}$$

Bottom slope: 0.001,
Manning roughness (earthen channel) (see Table 5.2), $n = 0.025$
Side slope: earthen channel, $m = 3/1$
Determine the desired depth

Solution:

Assume $B = 0.5$ m
Using Manning equation:

$$Q = \frac{1}{n} * A * R^{2/3} * S^{0.5}$$

$$1 = \frac{1}{0.025} * \left(0.5y + my^2\right) * \left(\frac{0.5Y + my^2}{0.5 + 2y * \sqrt{1 + m^2}}\right)^{2/3} * \left(\sqrt{0.001}\right)$$

$$\text{Func.} = 1 - \frac{1}{0.025} * \left(0.5y + my^2\right) * \left(\frac{0.5Y + my^2}{0.5 + 2y * \sqrt{1 + m^2}}\right)^{2/3} * \left(\sqrt{0.001}\right)$$

Solving the function iteratively gives $Y = 0.65$ m
Freeboard from Table 5.5 for 1 m³/s is 0.5 m.
and
$V = Q/A = 0.624$ m/s, which is an acceptable velocity

Example 5.9: Economic Design

The design of open channel should be functional (carry the flow with proper velocity, side slope and freeboard), but at the minimum feasible cost. The following example shows how to achieve that.
The canal is concrete with best side slope of 1/1.

$Q = 1.53$ m³/s
$S_o = 0.001$

Concrete cost: \$80/m³
$n = 0.013$ (concrete lined channel)

Solution:

To design this channel to be functional while keeping the cost at a minimum, we write the following objective functions and constraints:
Cost Function (objective function) = (P*th*1)*\$/m³ (cost per meter length of channel)
th: thickness of concrete channel which is typically 4 inches (0.102 m)

$$P : \text{wetted perimeter} = b + 2y\sqrt{1+m^2}$$

Therefore,

$$\text{Cost Function (objective function)} = \left(b + 2y\sqrt{1+m^2}\right)\text{th} * 1 * \$ / m^3$$

Constraints:

$Q = 1.53$ m³/s
$B > 0.0$
$Y > 0.0$
Using an optimization program such as Solver, B, Y and V are calculated as:
$B = 0.66$ m
$Y = 0.80$ m
$V = 1.32$ m/s
$m = 1$
And the cost is: \$23.8/m of length of channel

Example 5.10: Design of a triangular unlined channel

Planned $Q = 1$ m³/s
Bottom slope: 0.001
Manning roughness (earthen channel) Table 5.2, $n = 0.025$
Side slope: Earthen channel, $m = 3/1$
Determine the desired depth

Solution:

In triangular channel, $B = 0.0$
Using Manning equation:

$$Q = \frac{1}{n} * A * R^{2/3} * S^{0.5}$$

$$1 = \frac{1}{0.025} * (B.y + my^2) * \left(\frac{B.Y + my^2}{B + 2y * \sqrt{1 + m^2}} \right)^{2/3} * \left(\sqrt{0.001} \right)$$

$$\text{Func.} = 1 - \frac{1}{0.025} * (3y^2) * \left(\frac{3y^2}{2y * \sqrt{1 + 9}} \right)^{2/3} * \left(\sqrt{0.001} \right) = 1 - 2.3 * y^{2.67} = 0.0$$

By iterative solution,

$$Y = 0.73$$

and
Freeboard from Table 5.5 for 1 m³/s is 0.5 m
and
$V = Q/A = 0.626$ m/s, which is an acceptable velocity for earthen channel

Example 5.11: Design of circular channel

Calculate the diameter of a concrete storm sewer for the following condition:

$Q = 1.0$ m³/s
$n = 0.013$ from Table 5.2
$S = 0.001$

Assume channel is 70% full, $d/D = 0.7$.
Determine the required diameter D
From Figure 5.14, Q/Q-full for d/D of 0.7 is equal to 0.84
and Q-full = 1.0/0.84 = 1.18 m³/s
Using Manning:

$$Q \ \text{full} = \frac{1}{0.013} * \left(\frac{\pi D^2}{4} \right) * \left(\frac{D}{4} \right)^{2/3} * 0.001^{0.5}$$

and from here $D = 1.18$ m
V-full = Q-Full/A-full = 1.18/1.093 = 1.08 m/s
V/V-full = 1.12 from Figure 5.14
and
$V = 1.08*1.12 = 1.21$ m/s, which is acceptable for concrete channel.

Example 5.12: Flow rate calculation

Calculate the flow rate for a concrete circular sewer line with the following characteristics:
$S_o = 0.001$
$D = 1.2$ m
Manning $n = 0.013$
The pipe is 75% full,
$d/D = 0.75$
From Figure 5.14,

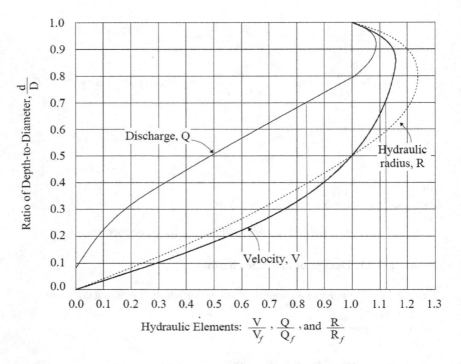

FIGURE 5.14 Dimensionless diagram for circular channel. This figure gives velocity and flow rates ratios as function of depth ratio.

Q/Q-full = 0.95

$$Q\text{-full} = \frac{1}{0.013} * \left(\frac{\pi D^2}{4}\right) * \left(\frac{D}{4}\right)^{2/3} * 0.001^{0.5} = 1.28 \ m^3/s$$

$Q/1.28 = 0.95$, $Q = 1.22 \ m^3/s$
V/V-full= 1.14 from Figure 5.3
Area-Full = $(3.14*1.2^2)/4 = 1.13 \ m^2$
V-full = $1.28/1.14 = 1.12$ m/s
$V = 1.14*1.12 = 1.277$ m/s

REFERENCES

Chow, V. T. (1959). *Open Channel Hydraulics*, McGraw Hill Publishing, New York.
Garcia, J. (1986). *Open Channel Hydraulics Manual*, Utah State University Publication.
Henderson, F. M. (1966). *Open Channel Flow*, McMillan Publishing, New York.
Simon, A. L. (1981). *Practical Hydraulics*, John Willey & Sons, New York.

6 Hydrology and Design of Culverts

6.1 HYDROLOGY AND RUNOFF

There are various methods to estimate peak runoff and volume of runoff from rain storms. The peak runoff depends on how the intensity of rainfall corresponds to the time of concentration of the watershed and the area of the watershed. Time of concentration is the time required for runoff to travel from the most remote point of the watershed to the outlet. One of the simplest methods of calculating peak runoff is the rational method developed by NRCS.

For storm sewers, most engineering offices in the United States use the rational design method which has been in use since 1886. This method, recommended by the Federal Highway Administration for roadside channels draining less than 200 acres, uses the equation

$$Q = C.i.A \qquad (6.1a)$$

where

Q = peak rate of runoff, in cfs
C = weighted runoff coefficient, expressing the ratio of rate of runoff to rate of rainfall (Table 6.1)
i = average intensity of rainfall, in inches per hour (for the selected frequency and for duration equal to the time of concentration)
A = drainage area, in acres, tributary to design point

This formula, although not dimensionally correct, gives numerically correct results since 1 cfs runoff equals 1.008 in. per acre. Table 6.1 shows some typical C values for various watersheds.

6.1.1 Watershed Characteristics

Some of the watershed characteristics that influence the amount and rate of runoff are:

1. Area and shape
2. Steepness and length of slopes
3. type and extent of vegetation or cultivation
4. Condition of surface: dry, saturated, frozen-pervious or impervious soil
5. Number, arrangement and condition of drainage channels on the watershed

DOI: 10.1201/9781003287537-6

TABLE 6.1
Values of Runoff Coefficient (C) for Various Conditions

For all watertight roof surfaces	0.75 to 0.95
For asphalt pavement	0.7 to 0.9
For concrete pavement	0.80 to 0.90
Residential – single family area	0.3 to 0.50
For impervious soils (heavy)	0.40 to 0.65
For impervious soils, with turf	0.30 to 0.55
For slightly pervious soils	0.15 to 0.40
For slightly pervious soils, with turf	0.10 to 0.30
For moderately pervious soils	0.05 to 0.20
For moderately pervious soils, with turf	0.0 to 0.10

The changes of land use during the lifetime of a drainage structure should be considered in evaluating runoff characteristics. Where the drainage area is composed of several types of ground cover, the runoff should be weighted according to the area of the melt type of cover present.

6.1.2 TIME OF CONCENTRATION

An important factor is the time required for runoff to reach from the remotest part of the drainage area to the design point. This is known as the time of concentration. It is used in the rational design method but must be clearly understood to avoid misapplication of the method and its intended purpose.

A minimum time of 5 minutes is recommended by the Federal Highway Administration.

6.1.3 DRAINAGE AREA

The drainage area can be measured on a topographic map or determined in the field by estimation, paring, aerial photos or a survey map.

Example 6.1

Find the discharge of a 10-year frequency rainfall at the outlet of a watershed in Miami, FL, with total area of 200 acres and time of concentration of 60 minutes. The weighted $C = 0.35$.

From Figure 6.1, the rainfall intensity for 10-year storm with rain of 60 min. is 3.5 in/hr. The runoff is calculated from Equation (6.1a) as:

$$Q = C.i.A = 0.35 \times 3.5 \times 200 = 245\,\text{cfs}$$

The volume of runoff can be calculated using curve number method. Curve number is an index of the soil's ability to store water. Curve number is a function of soil infiltration capacity and land cover (land use). Table 6.2 shows curve

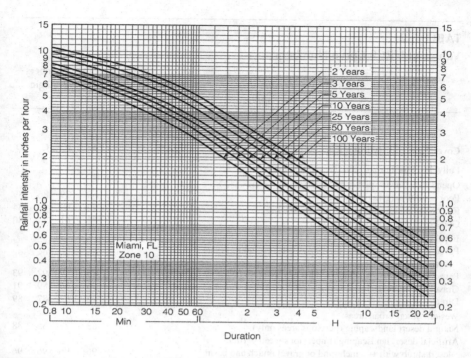

FIGURE 6.1 Intensity-duration-frequency (IDF) curves for Miami, FL.

numbers for soil hydrologic groups, which are a function of soil basic intake rate (Table 6.3). If the soil is made up of various layers, then the intake rate of the most restricting layer is used. The equations to be used to calculate volume of runoff are shown below.

$$S = \frac{1000}{CN} - 10 \qquad (6.1b)$$

$$Q = \frac{(P - 0.2S)^2}{P + 0.8S} \qquad (6.1c)$$

In which:

CN: is curve number
S: represents the soil's storability
P: precipitation, inches
Q: is runoff, inches

TABLE 6.2
Values of NRCS Curve Numbers

Cover Description		Curve Numbers for Hydrologic Group			
Cover Type and Hydrologic Condition	Average percent Impervious Area %	A	B	C	D
Fully developed urban areas (vegetation established)					
Open space (lawns, parks, golf courses, cemeteries, etc.)					
Poor condition (grass cover < 50%)		68	79	86	89
Fair condition (grass cover < 50% to 75%)		49	69	79	84
Good condition (grass cover > 75%)		39	61	74	80
Impervious areas:					
Paved parking lots, rooves, driveways, etc. (excluding rights of way)		98	98	98	98
Paved, open ditches (including rights of way)		83	89	92	93
Gravel (including rights of way)		76	85	89	91
Dirt (including rights of way)		72	82	87	89
Western desert urban areas:					
Natural desert landscaping (pervious areas only)		63	77	85	88
Artificial desert landscaping (impervious weed barrier, desert shrub with 1–2 inch sand or gravel mulch and basin borders)		96	96	96	96
Urban districts:					
Commercial and business	85	89	92	94	95
Industrial	72	81	88	91	93
Residential districts by average lot size:					
1/8 acre or less (town houses)	65	77	85	90	92
1/4 acre	38	61	75	83	87
1/3 acre	30	57	72	81	86
1/2 acre	25	54	70	80	85
1 acre	20	51	68	79	84
2 acre	12	46	65	77	82
(pervious areas, no vegetation)		77	86	91	94

TABLE 6.3
Soil Hydrologic Group

Soil Minimum Intake Rate, mm/h	Group
>7.6	A
>3.8 to 7.6	B
1.3–3.8	C
<1.3	D

Example 6.2

In Example 6.1, if the total precipitation is 6.5 inches and curve number is 74, what is the volume of runoff in acres-ft?

From Equations (6.1b) and (6.1c), the volume is calculated as follows:

$$S = \frac{1000}{74} - 10 = 3.5$$

And

$$Q = \frac{(6.5 - 0.2 * 3.5)^2}{6.5 + 0.8 * 3.5} = 3.6 \text{ inches}$$

and the volume of runoff is

$$\text{Volume} = 200(3.6)/12 = 60 \text{ Acre-ft}$$

6.2 CULVERT DESIGN

Before designing culverts, storm sewers and other drainage structures, it is important to consider the design of ditches, gutters, chutes, and other channels leading to these structures.

Rainfall and runoff, once calculated, are followed by design of suitable channels to handle the peak discharge with minimum erosion, maintenance and hazard to traffic.

The AASHO Policy on Geometric Design of Rural Highways recommends:

> Where terrain permits, roadside drainage channels built in earth should have side slope not steeper than 4:1 (horizontal to vertical), and a rounded bottom at least 4 ft wide. (Minimum depth 1 ft to 3 ft). Dimensions can be varied by the use of different types of channel lining.

The complete Manning formula in English units is:

$$Q = A \frac{1.486}{n} R^{2/3} S^{1/2} \qquad (6.2)$$

where

A = cross-sectional area of flow in sq. ft
S = slope in ft per ft
R = Hydraulic radius in ft = A/P
P = Wetted perimeter
n = Coefficient of roughness (Table 6.4)

TABLE 6.4
Roughness Values "*n*" for Various Type of Lining

Type of Lining	*n* (Manning)
Ordinary earth, smoothly graded	0.02
Sod depth of flow over 6 inches	0.04
Sod depth of flow below 6 inches	0.06
A rip-rap, rough	0.04
Concrete paved gutter	0.016

6.2.1 HYDRAULICS OF CULVERTS

The design process for a culvert consists of (1) collecting field data, (2) compiling facts about the roadway, and (3) making a reasonable estimate of flood flow for a chosen frequency. The fourth step is to design an economical culvert to handle the flow (including debris) with minimum damage to the highway, street, railway or adjacent property (Figure 6.2).

6.2.2 WHAT MAKES A GOOD CULVERT

1. The culvert, appurtenant entrance and outlet structures should properly take care of water, bed-load and floating debris at all stages of flow.
2. It should cause no unnecessary or excessive property damage.
3. It should provide for transportation of material without detrimental change in flow pattern above and below the structure.
4. It should be designed so that future channel and highway improvements can be made without too much loss or difficulty.
5. It should be designed to function properly after fill has caused settlement.
6. It should not cause objectionable stagnant pools in which mosquitos may breed.
7. It should be designed to accommodate increased runoff occasioned by anticipated land improvement.
8. It should be economical to build, hydraulically adequate to handle, designed to discharge, structurally durable and easy to maintain.
9. It should be designed to avoid excessive ponding at the entrance which may cause property damage, accumulation of drift, culvert clogging, saturation of fills or detrimental upstream deposit of debris.
10. Entrance structures should be designed to screen out materials which will not pass through the culvert, reduce entrance loss to a minimum, make use of the velocity of approach insofar as practicable, and by use of transitions and increased slopes, as necessary, facilitate channel flow entering the culvert.
11. The design of culvert and outlet should be effective in re-establishing tolerable non-erosive channel flow within the right of way or within a reasonably short distance below the culvert.

FIGURE 6.2   Pipe concrete culvert for storm runoff under a road, Las Cruces, NM.

12. The outlet should be designed to resist undermining and washout.
13. Culvert dissipaters, if used, should be simple, easy to build, economical and
    reasonably self-cleaning.

6.3   TYPES OF CULVERTS

Culverts are divided into two categories. These are inlet control and outlet control as
shown in Figures 6.3–6.5.

6.3.1   CULVERTS WITH INLET CONTROL

Inlet control is one of the two major types of culvert flow. An inlet-control culvert
with un-submerged culvert inlet is preferable to a submerged one.

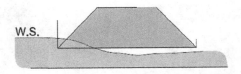

FIGURE 6.3   Culvert with inlet control.

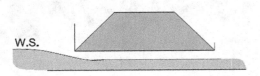

FIGURE 6.4   Culvert with inlet control.

FIGURE 6.5   Outlet-controlled culvert, $(H = H_1 - H_2)$.

6.3.2   CULVERTS WITH OUTLET CONTROL

In the outlet-control culvert (Figure 6.5), the flow is controlled by both an inlet hydraulic head and a tail water head.

6.4   HYDRAULIC CONDITIONS AND DEFINITIONS

Conventional culverts considered here are circular pipes and pipe-arches, with a uniform barrel cross-section throughout.

6.4.1   INLET CONTROL

Under inlet control, (1) the cross-sectional area of the barrel, (2) the inlet configuration or geometry and (3) the amount of headwater or ponding are of primary importance (Figures 6.3 and 6.4).

6.4.2   OUTLET CONTROL

Outlet control involves the additional consideration of the tail water in the outlet channel, and the slope, roughness and length of the barrel (Figure 6.5).

6.5 HYDRAULICS OF OUTLET-CONTROLLED CULVERT

Head (H): The head is energy required to pass a given quantity of water through a culvert flowing in outlet control (with barrel full). It is made up of (1) velocity head $H_v = \dfrac{V^2}{2g}$, (2) an entrance loss H_e, and (3) a friction loss H_f. This energy is obtained from ponding at entrance and slope of pipe, and is expressed in the following form:

$$H = H_v + H_e + H_f, \tag{6.3}$$

H = Total head loss
H_v = Exit loss
H_e = Entrance loss
H_f = Friction loss

Entrance loss (H_e) depends upon the geometry of the inlet edge. This loss is expressed as a coefficient K_e multiplied by the barrel velocity head or

$$H_e = k_e \frac{V^2}{2g} \tag{6.4}$$

Friction loss (H_f) is the energy required to overcome the roughness of the culvert barrel and is expressed in the following equation:

$$H_f = \frac{\left(26n^2 L\right)}{R^{1.33}} \frac{V^2}{2g} \tag{6.5}$$

where:

n = Manning's friction factor (see Table 6.2)
L = length of culvert barrel (ft)

TABLE 6.5
Entrance Loss Coefficient (K_e) for Corrugated Metal Pipe or Pipe Arch

Inlet End of Culvert	Coefficient K_e
Projecting from fill (no headwall)	0.60
Headwall, or headwall and wingwall square-edge	0.50
Mitered (beveled) to conform to fill slope	0.70
End section conforming to fill slope	0.50
Headwall, rounded edge	0.20
Beveled ring	0.25

V = mean velocity of flow in barrel (ft/sec)

g = acceleration of gravity, 32.2 (ft/sec³)

R = Hydraulic radius or $\dfrac{A}{WP}$ (ft)

Combining Equations (6.4) and (6.5) we get the full flow culvert equation for the English system:

$$H = H_1 - H_2 = \left(1 + K_e + \frac{26n^2 L}{R^{1.33}}\right)\frac{V^2}{2g} \tag{6.6}$$

For the metric system, Equation (6.7) is used:

$$H = \left(1 + K_e + \frac{16.62n^2 L}{R^{1.33}}\right)\frac{V^2}{2g}, \tag{6.7}$$

6.6   HYDRAULICS OF INLET-CONTROLLED CULVERT

The flow through an inlet-controlled culvert can be calculated using an orifice equation at the inlet:

$$Q = C.A.\sqrt{2gH}, \tag{6.8}$$

where

C is orifice coefficient = 0.6

A = cross-sectional area of inlet

g = acceleration of gravity

H = head above the centroid of inlet

Example 6.3: Design of an outlet-controlled culvert

We want to design a concrete pipe culvert with headwall, and length of 120 ft to carry maximum flow of 30 cfs. The water levels upstream and downstream of the culvert are at 1.8 ft and 1.05 ft respectively.

A.   Determine the required diameter of the culvert.

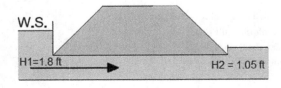

Using Equation (6.6), we can set up a function and solve for D as shown below:

$$\text{Function} = H - \left(1 + K_e + \frac{26n^2L}{R^{1.33}}\right)\frac{V^2}{2g} = 0.0$$

Start with assuming D of 2 ft, and $K_e = 0.5$ and $n = 0.016$. Using an iterative approach,

D, ft	V, ft/s	H, ft	Function
2	9.55	0.75	−4.20
3	4.246	0.75	0.002

Therefore a culvert diameter of 3 ft will work, provided it is not controlled by the inlet.

Using Equation (6.8), the capacity of the inlet can be calculated as follows:

Assuming the center of the pipe culvert is 0.9 ft. below the water surface

$$Q = C.A.\sqrt{2gH}$$

$$Q = 0.6\left(\frac{3.14 \times 3^2}{4}\right)\sqrt{2g(0.9)} = 32.2\,\text{cfs}$$

Since Q of the inlet (32.2 cfs) is greater than the design capacity (30 cfs), the culvert is functional.

7 Simple Flow Measurement Structures

This chapter was extracted and modified from a previous report by Davis and Samani (2016).

7.1 INTRODUCTION

With increasing demands from agriculture, municipal needs, industry, and recreational use, there is a need for water users to conserve, use, and share water wisely. Measuring water in open channels is an important first step toward water conservation. The measurement of the amount of water delivered and received by users will help ensure that each gets their fair share, establish a more equitable distribution

DOI: 10.1201/9781003287537-7

of available water and promote conservation. Ever since the development of the Parshall Flume (Parshall, 1926), continuous improvements have been made to simplify, reduce costs, increase accuracy, and reduce head losses in open-channel flumes (Hagar, 1988, Skogerboe et al., 1967, 1972, Replogle, 1975, Samani et al., 1991, Samani and Magallanez, 1993, 2000).

As we study the importance of water use and management, the importance of knowing not only the quality, but quantity of water becomes more apparent. Traditional flumes and weirs are practical in many scenarios, allowing for accurate flow measurement. However, with low-gradient, high-sediment channels found in many parts of the world, coupled with the financial constraints and lack of technical expertise, there is a need for a simple, cost-effective way to measure discharge. Simple flumes do not require extensive upstream transition, reducing the amount of materials needed and the construction costs. They produce minimal head loss and are self-flushing and do not accumulate the sediment often found in weirs or even in traditional weirs in low-gradient conditions. Simple flumes have been found to be an attractive and accurate option to consider when measuring flow.

7.2   PRINCIPLES OF FLOW MEASUREMENTS IN OPEN CHANNELS

To understand flow measurement, it is necessary to understand how water flows. Water flow is based on the principle of energy. Two types of energy govern flow in open channels: global and specific energy. According to the principle of specific energy, minimum specific energy occurs when the Froude number (Fr) is equal to 1 (see Figure 7.1)

$$\mathrm{Fr}^2 = \frac{Q^2 T}{g\left(A^3\right)} = 1 \qquad (7.1)$$

where:

  Q = flowrate
  A = cross-sectional area
  T = top width of the flow
  g = acceleration of gravity

This condition of minimum energy also produces a specific flow called critical flow. As can be seen in Equation (7.1), when critical flow occurs, the flow rate (Q) can be calculated through the cross-section area and top width of the flow. To

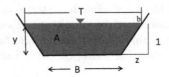

FIGURE 7.1   Typical channel dimensions shown in a channel cross-section.

accurately measure flow in an open channel, it is necessary to create a critical flow condition, allowing flow to be measured by simply measuring the flow depth (yielding values for both T and A).

There are three ways of creating critical flow in open channels: 1. raising the bottom of the channel; 2. lowering the bottom of the channel; 3. contracting the flow cross-section (narrowing the channel). Flow measurement devices strive to achieve this critical flow for various conditions thus providing an accurate means of measurement.

7.3 SIMPLIFIED FLUMES

Flumes are specially shaped open-channel flow sections that are created to constrict flow in a canal or ditch and induce critical flow, allowing the flow rate to be determined. They are constructed of metal, concrete, wood, or fiberglass and can measure water over a wide variety of flow ranges. Larger flumes are constructed on site, while smaller flumes can be prefabricated and installed later on site.

Simple flumes require relatively little head loss, can operate in flat ditches, are relatively insensitive to approach velocity, and are self-cleaning due to the high velocity of water passing through. Flumes must be leveled and, to provide accurate flow measurements, the flow entering the flume cannot contain waves or surges.

Critical flow can be created by contracting the cross-section of an existing canal without changing the existing canal dimensions. Simple flumes create a small section of critical flow before water returns to its previous energy state after a hydraulic jump has occurred. Minimal head loss is required and no extended inflow or outflow transition is needed. These flumes reduce the cost, calculations, and head loss while minimizing the materials needed to measure flow. As canals come in various shapes and sizes, various short-throated flumes have been developed to measure the respective flows. The three simple flume types that have been developed are circular, rectangular (S–M), and trapezoidal flumes (Figures 7.2, 7.11, and 7.20).

When the flow cross-section is contracted to create a critical flow, the discharge rate depends on the upstream energy (H), the width of the critical section (B_c), and the acceleration of gravity (g) as seen in Figures 7.4 and 7.5. Therefore flow can be described as a function of H, B_c, and g as:

$$Q = \text{function}\left(H, B_c, g\right)$$

Gravity remains a constant, and the critical section (B_c) can be calculated for each channel type, leaving the upstream energy (H) as the only variable needed to calculate flow.

According to Samani (2017), the function of Q was found to be:

$$Q = a\left(B_c^{2.5}\right)\left(g^{0.5}\right)\left(\frac{H}{B_c}\right)^b \tag{7.2}$$

where:
Q = flowrate, ft³/s or m³/s
B_c = critical cross-section width, ft or m

TABLE 7.1

Coefficients for Various Short-Throated Flumes Fitting Equation (7.2)

Flume Type	a	b
Circular	0.421	2.31
Rectangular	0.701	1.59
Trapezoidal	0.226	1.51

g = gravity, ft/s^2 or m/s^2
H = upstream height of water at flume, ft or m

a and b are empirical coefficients that are determined experimentally using the laboratory scale models shown in Table 7.1 (Samani, 2017). These values are the same when using both imperial and metric measuring systems.

Once the dimensions of the channel and flume are known, B_c becomes a function of H, making the upstream height of water (H), the only variable needed to determine discharge.

7.4   CIRCULAR FLUMES (SAMANI ET AL., 1991)

Pipes are often used to drain or convey water, and partially full pipes are considered open channels. The flume created in these circular channels consists of a vertical pipe installed inside a horizontal circular pipe to create critical flow. Cross-sectional and profile views of the circular flume are shown in Figures 7.3 and 7.4. The head or height of the flow would be measured at the center of the upstream side of the vertical column as indicated by the ruler in Figure 7.2.

Inserting the coefficients for circular flumes from Table 7.1 into Equation (7.2), the equation for discharge using a circular flume is:

$$Q = 0.421 * B_c^{2.5} * g^{0.5} * \left(\frac{H}{B_c}\right)^{2.31} \tag{7.3}$$

where:

Q = discharge, ft^3/s or m^3/s
H = head, ft or m
g = gravity, ft/s^2 or m/s^2
B_c = width of channel at critical cross section in feet or meters. Given by:

$$B_c = D - d \tag{7.4}$$

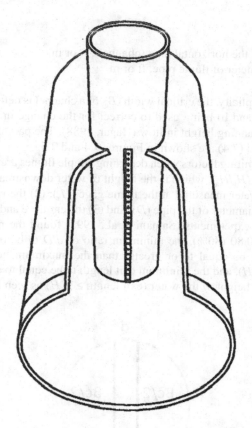

FIGURE 7.2   Cut-out view of a circular flume from the upstream side (Samani et al., 1991).

FIGURE 7.3   Circular flume in southern New Mexico irrigation canal, located at GPS coordinates: 32.359167, 106.784722.

where:

 D = diameter of the horizontal pipe (channel), ft or m
 d = outside diameter of flume pipe, ft or m

For the sake of simplicity, the critical width (B_c) of a channel is defined by $D - d$, with the coefficients (a and b) being used to correct for the change in critical width that occurs with the changing height in flow (Hagar, 1988). The parameters described in Equations (7.3) and (7.4) are shown in Figures 7.4 and 7.5.

 The principle limiting factors when designing simple flumes are: (1) the maximum submergence ratio (H_b/H_a), which is the height of water downstream (H_b) divided by the height of the water measured at the flume gage (H_a); (2) the ratio of the vertical column (d) to the diameter of the pipe (D); and (3) the entrance and exit lengths of the flume. Laboratory experiments (Samani et al., 1991) found the maximum submergence limit to be 0.80 (80%), the minimum ratio of d/D to be 0.25, the minimum entrance length to be equal to or greater than the maximum height of the water (entrance length $\geq H$), and the minimum exit length to be equal to or greater than one half the maximum height of the water (exit length $\geq \frac{1}{2} H$) as seen in Figure 7.5.

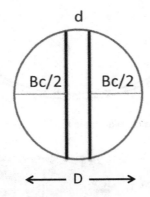

FIGURE 7.4   Cross-sectional view of a circular flume.

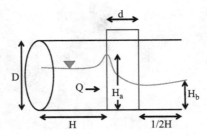

FIGURE 7.5   Critical design parameters, where $H_b/H_a \leq 0.8$ and $d/D \geq 0.25$.

FIGURE 7.6 Circular flume installed in a vineyard in Australia.

One advantage of the circular flume is that it functions similar to a v-notch weir, having a large exponent (2.31) thus being able to measure a large range of flows. This makes the circular flume ideal for variable flow conditions. The circular flume has been adopted for use in the United States and various other countries. Some of their applications can be seen in Figures 7.3 and 7.6.

7.4.1 DESIGN EXAMPLE OF A CIRCULAR FLUME

The following is an example of how to design and calibrate a circular flume for a given channel, known flow rate, and known normal maximum water depth.

Example 7.1

Existing circular canal with a diameter: 1.2 ft
  Maximum flow rate: 1.0 ft³/s
  Normal depth at maximum flow: 0.57 ft
  We first assume a diameter of the interior vertical flume, generally in pipe sizes that are readily available. For this example, we will start with the minimum allowable ratio of 0.3 ft ($d/D = 0.25$, $d = 0.25 * 1.2$ ft).

$$B_c = D - d = 1.2\,\text{ft} - 0.3\,\text{ft} = 0.9\,\text{ft} \tag{7.4}$$

Using Equation (7.3):

$$1.0 = 0.421 * (0.9)^{2.5} * \sqrt{32.2} * \left(\frac{H}{0.9}\right)^{2.31}$$

$$H = 0.692 \text{ ft}$$

Submergence ratio: 0.57/0.692 = 0.824, which exceeds the maximum submergence ratio of 0.8.

Therefore, we can increase the diameter of the column to $d = 0.60$ ft, $D - d = 0.6$ ft, and recalculate H.

$$H = 0.715 \text{ ft}$$

Submergence ratio: 0.57/0.715 = 0.80, which meets the submergence requirement.

The entrance length would need to be at least 0.715 ft and the exit length would need to be at least 0.36 ft. This flume would require a vertical pipe of outside diameter 0.6 ft or greater and would measure flows up to 1.0 cfs (448.4 gpm).

7.5   S–M FLUME (SAMANI AND MAGALLANEZ, 2000)

Rectangular flumes (S–M flume) can be constructed and used in irregular channels (Figures 7.7 and 7.12). They are the easiest flume type to construct. The contraction of the water and the critical flow section can be made to pass through the center of the

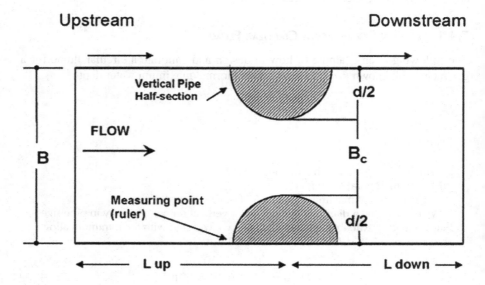

FIGURE 7.7   Plan view and common dimensions of S–M rectangular flume indicating the point of measurement with design parameter of $d/B \geq 0.4$ (Samani, 2005).

channel, making it ideal for flow with high sediment loads. This flume was first created, tested, and named by Samani and Magallanez (2000) and is named the S–M flume.

The contraction is made by cutting a pipe in half and placing a half-pipe on each side of the channel opposite each other with the gage set on the upstream side of the flume as seen in Figure 7.7. This creates critical flow between the two half pipes while flushing sediment and debris through the flume.

Inserting the coefficients from Table 7.1 into Equation (7.2), the stage-discharge equation for the S–M flume is given by:

$$Q = 0.701 * B_c^{2.5} * g^{0.5} * \left(\frac{H}{B_c}\right)^{1.59} \tag{7.4}$$

where:

Q = discharge, ft³/s or m³/s
H = height of water, ft or m
g = gravity, ft/s² or m/s²
B_c = width of channel at critical cross section in feet or meters. Given by:

$$B_c = B - d \tag{7.6}$$

where:

B = width of the base of the channel, ft or m
d = diameter of the half pipe, ft or m

The principle limiting factor when designing simple flumes is the maximum submergence ratio (H_b/H_a), which is the height of water downstream (H_b) divided by the height of the water measured at the flume gage (H_a). Laboratory experiments (Samani, 2000) found this maximum limit to be 0.80 (80%).

The ratio of the diameter of the pipe (d) to the width of the rectangular channel (B) is the contraction ratio. This ratio (d/B) should be greater than 0.40 (40%) or conversely $B_c/B \leq 0.6$ (60%), thus ensuring critical flow and allowing the discharge to be accurately measured.

The minimum entrance length needs to be equal to or greater than the maximum height of the water (entrance length $\geq H$), and the minimum exit length to be equal to or greater than one half the maximum height of the water (exit length $\geq \frac{1}{2} H$) as seen in Figure 7.8.

Although rectangular, S–M flumes are very universal and can easily be applied to irregular (Figures 7.12 and 7.13) or trapezoidal-shaped channels (as shown in Figure 7.9) by filling in the sides of the channel.

The S–M flume can also be used to measure flow in parallel at varying heights and can be placed in high sediment- or debris-prone areas such as flood channels (Figure 7.10). This would allow the middle flume to measure discharge at low flow rates

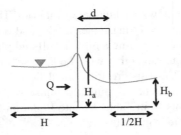

FIGURE 7.8 Diagram indicating critical design parameters, where $H_b/H_a \leq 0.8$.

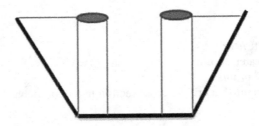

FIGURE 7.9 Cross-sectional view of S–M flume retrofitted into a trapezoidal channel.

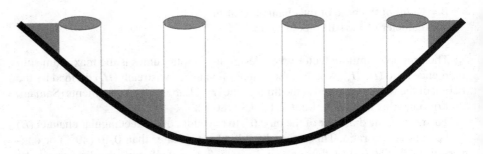

FIGURE 7.10 Multi-stage diagram of an S–M flume which could measure at both low and high discharge rates while allowing sediment to pass.

while allowing debris to pass through. At higher flow rates all three flumes would measure flow through each section which could be added together to obtain the total flow. In these higher flow scenarios, the middle flume would act as a sluice gate, flushing sediment and debris through the flume.

The S–M flume can also be installed at turnouts as shown in Figure 7.11 and with the flume functioning similarly to a large turnout. The induced critical flow and subsequent hydraulic jump dissipate the energy of the water prior to entering the field.

FIGURE 7.11 An S–M flume installed in a turnout in a pecan farm in La Union, NM (Samani, 2005).

FIGURE 7.12 A pre-fabricated S–M flume installed in an unlined irregular channel (*acequia*) in Santa Fe, NM.

FIGURE 7.13   S–M rectangular flume installed in an unlined ditch in southern New Mexico.

7.5.1   Design Example of an S–M Flume

The following is an example of how to design and calibrate an S–M flume for a given rectangular channel, with known flow rate, and known normal maximum water depth. Additional solutions are provided in the case that a large enough pipe is not available. In this case the base of the flume could be elevated or the sides of the channel could be contracted.

Example 7.2

Existing rectangular canal with base width: 1.2 ft
Maximum flow rate: 1.0 ft³/s
   Normal depth at maximum flow: 0.42 ft
   We first assume a diameter of the interior vertical flume, generally in pipe sizes that are readily available. For S–M flumes we also need to meet the minimum contraction ratio of 0.40. For this example, we will assume an outside diameter of 0.3 ft.
   This gives a contraction ratio of 0.3/1.2 = 0.25, which is below the minimum contraction ratio threshold.
   Increasing d to 0.5 ft increases our contraction ration to 0.42 which meets the requirement.

$$B - d = 1.2\,\text{ft} - 0.5\,\text{ft} = 0.7\,\text{ft} \tag{7.6}$$

Using Equation (7.5):

$$1.0 = 0.701*(0.7)^{2.5}*\sqrt{32.2}*\left(\frac{H}{0.7}\right)^{1.59}$$

$$H = 0.515 \text{ ft}$$

Submergence ratio: 0.42/0.515 = 0.815, this still exceeds the maximum submergence ratio of 0.8.

Therefore, we can increase the diameter of the pipe to $d = 0.6$ ft, $D - d = 1.2$ ft – 0.6 ft = 0.6 ft, and recalculate H (Equation (7.5)).

$$H = 0.562 \text{ ft}$$

Submergence ratio: 0.42/0.562 = 0.75 which meets the submergence requirement.

The entrance length would be a minimum of 0.562 feet, the exit length a minimum of 0.281 feet. This flume would require a vertical pipe with outside diameter 0.6 ft or greater and would measure flows up to 1.0 cfs (448.4 gpm).

7.5.2 Alternate Solutions

Another solution would be to elevate the base of the channel slightly instead of increasing the pipe diameter from 0.3 ft to 0.6 ft.

Dividing the normal depth (0.42 ft) by the submergence limit (0.80), we obtain 0.42 ft/0.80 = 0.525 ft.

The upstream height must be at least 0.525 ft.

Using the first pipe of 0.3 ft, we obtain an upstream height of 0.515 ft.

Subtracting the upstream height (0.515 ft) from the required height (0.525 ft); 0.525 ft – 0.515 ft = 0.010 ft.

By raising the base of the channel under the flume by 0.010 feet we would be able to use the vertical pipe diameter of 0.3 ft and meet the required submergence ratio.

An additional solution would be to further contract the sides of the flume to create the allowable submergence ratio. We saw in the first solution that a half-pipe with diameter of 0.6 feet provided for proper submergence.

Supposing that we only have a pipe of diameter 0.3 feet, we could instead fill in each side of the channel with 0.3 feet of concrete or other material.

This would create the same critical width (B_c) of 0.6 feet (1.2 ft – 0.6 ft), and give the same gage height of 0.562 feet found in the first solution which has been verified to work.

7.6 TRAPEZOIDAL FLUMES

Most irrigation canals are trapezoidal in shape due to the high hydraulic efficiency of trapezoidal channels. As such, it becomes increasingly relevant and convenient to design a low-cost, easily installed flume that can be used to measure flow in trapezoidal canals (Samani and Magallanez, 1992, Badar and Ghare, 2012). The flume was designed and installed similarly to how circular flumes are installed, with a vertical pipe in the middle of a channel as seen in Figures 7.14, and 7.15.

Inserting the coefficients from Table 7.1 into Equation (7.2), the equation for flow for this trapezoidal flume is:

$$Q = 0.226 * \left(B_c\right)^{2.5} * g^{0.5} * \left(\frac{H}{B_c}\right)^{1.51} \tag{7.5}$$

FIGURE 7.14 Trapezoidal flume in southern New Mexico Irrigation canal located at GPS coordinates: 32.391389, 106.846111.

FIGURE 7.15 Close-up (upstream side) of previous flume, with visible rust patterns indicating maximum water height and critical flow profile.

where:

Q = discharge, ft³/s or m³/s
H = height of water, ft or m
g = gravity, ft/s² or m/s²
B_c = width of channel at critical cross section in feet or meters. Given by:

$$B_c = B + 2mH - d \qquad (7.6)$$

where:

B = width of the base of the channel, ft or m
m = side slope of channel, unitless (see Figure 7.20)
H = height of water, ft or m
d = diameter of the vertical pipe, ft or m

The principal limiting factor when designing simple flumes is the maximum submergence ratio (H_b/H_a), which is the height of water downstream (H_b) divided by the height of the water measured at the flume gage (H_a). Laboratory experiments (Samani, 1993) found this maximum limit to be 0.80 (80%).

Laboratory experiments also shown that the ratio of the column diameter (d) to the canal bottom width (B), (d/B), should be 75% or greater for critical flow to occur (Figures 7.16 and 7.17).

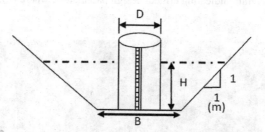

FIGURE 7.16 Cross-sectional view of a trapezoidal flume with common dimensions.

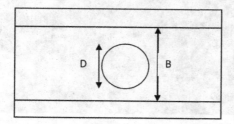

FIGURE 7.17 Plan view of a trapezoidal flume indicating D, diameter of flume and B, canal bottom width with design parameter $D/B \geq 0.75$.

The minimum entrance length to be equal to or greater than the maximum height of the water (entrance length ≥ H), and the minimum exit length to be equal to or greater than one half the maximum height of the water (exit length ≥ ½ H) as seen in Figure 7.18.

Downstream and upstream flow conditions during operation are shown in Figures 7.19 and 7.20.

Figure 7.21 shows a trapezoidal flume installed in a concrete-lined canal in Dushanbah, Tajikstan.

7.6.1 DESIGN EXAMPLE OF A TRAPEZOIDAL FLUME

The following is an example of how to design and calibrate a trapezoidal flume for a given channel, with known flow rate, and known normal maximum water depth. An additional solution is provided in the case that a large enough pipe is not available. In this case the base of the flume could be elevated.

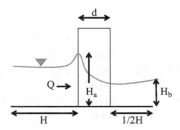

FIGURE 7.18 Diagram indicating critical design parameters, where $H_b/H_a \leq 0.8$.

FIGURE 7.19 Trapezoidal flume testing in the lab with critical flow.

FIGURE 7.20   Trapezoidal flume as seen from downstream with both visible flow depths.

FIGURE 7.21   Trapezoidal flume in use in Dushanbah, Tajikistan.

7.6.2 Alternate Solution

Another solution would be to elevate the base of the channel slightly instead of increasing the pipe diameter from 0.9 ft to 1.0 ft.

Example 7.3

Existing trapezoidal canal with base: 1.2 ft
Existing side slope: 1:1 (m:1)
   Maximum flow rate: 1.0 ft³/s
   Normal depth at maximum flow: 0.52 ft
   We first assume a diameter of the vertical column, generally in readily available pipe sizes. We also need to meet the requirement of the column width being a minimum of 75% of the base of the channel. For this example, we will assume an outside diameter of 0.9 ft.

$$B + 2*m*H - d = 1.2\,\text{ft} + (2*1*H) - 0.9\,\text{ft} = 2H + 0.3\,\text{ft}, \qquad (7.8)$$

Using Equation (7.5):

$$1.0 = 0.226*(2H+0.3)^{2.5}*\sqrt{32.2}*\left(\frac{H}{2H+0.3}\right)^{1.51}$$

$$H = 0.632\,\text{ft}$$

Submergence ratio: 0.52/0.632 = 0.823, which exceeds the maximum submergence ratio of 0.8.
   Therefore, we can increase the diameter of the column to $d = 1.00$ ft, $D - d + 2mH = 0.20 + 2H$ ft, and recalculate H which is,

$$H = 0.650\,\text{ft}$$

Submergence ratio: 0.52/0.650 = 0.80, which meets the submergence requirement.
   The entrance length would be a minimum of 0.650 feet and the exit length would be a minimum of 0.325 feet. This flume would require a vertical pipe of outside diameter 1.0 ft or greater and would measure flows up to 1.0 cfs (448.4 gpm).

Dividing the normal depth (0.52 ft) by the submergence limit (0.80), we obtain 0.52 ft/0.80 = 0.65 ft.
   Meaning the upstream height must be at least 0.65 ft.
   Using the first pipe of 0.9 ft, we obtained an upstream height of 0.632 ft.
   Subtracting the upstream height (0.632 ft) from the required height (0.65 ft); 0.65 ft − 0.632 ft = 0.018ft.
   By raising the base of the channel under the flume by 0.018 feet we would be able to use the vertical pipe diameter of 0.9 ft and meet the required submergence ratio.

7.7 GENERAL DESIGN RECOMMENDATIONS

When designing a flume:

- The length of the flume should be ≥1.5 times the maximum height of the flow.
- The vertical column should be placed a minimum distance equal to the height of the water level from the entrance and ½ the height of the water level from the exit.
- In all instances, the height of the water should be measured at the center of the upstream side of the vertical column.
- Oftentimes, due to turbulence the water level will be difficult to read. In such cases the water level should be taken as an average of the fluctuating height.
- A more desirable option to prevent error caused by fluctuating water levels at the column would be to use the column as a stilling well by drilling a hole in the base of the vertical column on the upstream side and measuring the water depth inside the column. This will allow water to enter the column with little turbulence on the inside.
- Rating tables can easily be constructed once a flume is in place to prevent the need for constant calculation.

REFERENCES

Badar, A. M. and Ghare, A. D. (2012). Development of Discharge Prediction Model for Trapezoidal Canals using Simple Portable Flume. *International Journal of Hydraulic Engineering* 1(5): 37–42.

Cadena, F. and Magallanez, H. (2005). Analytical Solution for Circular Gates as Flow Metering Structures. *Journal of Irrigation and Drainage Engineering ASCE* 131(5): 451–456.

Davis, S. and Samani, Z. (2016). *Flow Measurement Devices for open channels. Report submitted to New Mexico Water Resources Research Institute*, Las Cruces, New Mexico.

Hagar, W. H. (1988). Mobile Flume for Circular Channel. *Journal of Irrigation and Drainage Engineering ASCE* 114(3): 520–534.

Manning, R. (1891). On the Flow of Water in Open Channels and Pipes. *Transactions of the Institution of Civil Engineers of Ireland* 20: 161–207.

Parshall, R. L. (1926). The Improved Venturi Flume. *Transactions ASCE* 89: 841–880.

Replogle, J. A. (1975). Critical Flow Flumes with Complex Cross Section. *Proc. ASCE Irrg. Drain. Div. Spec. Conf., ASCE*, August 13–15, 336–338.

Samani, Z. (2017). Three Simple Flume for Open Channel. *Paper in Progress. Journal of Irrigation & Drainage Engineering* 143(6): 04017010.

Samani, Z., Jorat, S. and Yousaf, M. (1991). (Hydraulic Characteristics of a Circular Flume. *Journal of Irrigation and Drainage Engineering ASCE* 117(4): 559–567.

Samani, Z. and Magallanez, H. (1993). Measuring Water in Trapezoidal Canals. *Journal of Irrigation and Drainage Engineering ASCE* 119(4): 181–186.

Samani, Z. and Magallanez, H. (2000). Simple Flume for Flow Measurement in Open Channel. *Journal of Irrigation & Drainage Engineering ASCE* 126(2): 127–129.

Samani, Z., Magallanez, H. and Skaggs, R. (2005). A Simple Flow Measuring Device for Farms. Water Task Force. Report 3.

Skogerboe, G. V., Bennet, R. S. and Walker, W. R. (1972). Generalized Discharge Relations for Cutthroat Flumes. *Journal of Irrigation and Drainage Engineering ASCE* 98(4): 569–583.

Skogerboe, G. V., Hyatt, M. L., Anderson, R. K. and Eggleston, K. O. (1967). Design and Calibration of Submerged Open Channel Flow Measurement Structures: Part 3 – Cutthroat Flumes. Report WG31-4. Utah Water Research Laboratory, College of Engineering. Utah State University, Logan, UT.

Index

Note: Pages in *italics* refer to figures.

Printed in the United States
by Baker & Taylor Publisher Services

Printed in the United States
by Baker & Taylor Publisher Services